After a year or so as Acting Dean and Provost at UCSC, I took my first sabbatical and decided to spend some time traveling as well as considering what I'd do for contrast. What then occurred was probably more a mid-life crisis than a break from academia. One of my first research studies was to combine art and science in the design of a radio telescope that supposedly worked, gave indications of changes through sound output (SETI) and looked, I thought, quite elegant. I then gathered a robust and professional group of scientists, artists, and musicians to help make my dream a reality. Unfortunately, the construction, no matter the beauty and high-grade functionality of the end design, never took place due to arguments between state and federal controls of land use and, frankly, the amount of money for construction required that climbed about $1 million per year as delay after delay postponed construction.

Pages 6-50 in this book show each of the twenty-two finalists from about twice that many drawings that I initially made for you to see the state of my mind at the time (early eighties), and from there to the end, the final design with all of the details described in detail including original funding requests. Some of this material may require readers have some knowledge of mathematics, botany, archeology, geography, radio astronomy, and animal husbandry, but suffice it to say all of this is available on the Internet or the local library.

Who knows, however, someone may come across this book and decide that it's time the world have multipurpose artistic and scientific tools and make Pleiades a reality after the nearly forty years since its inception.

Who knows?

Pleiades

⌘

David Cope

Pleiades

By David Cope

Epoc Books
Printed in the United States of America
© David Cope 2017
All Rights Reserved.
Published 1983, 2017.

Acknowledgments

My sincere thanks go to my wife Mary Jane, without whose encouragement and patience this book could never have been completed, Keith Muscutt, Mark Primack, Peter Elsea, Michael Stamp, David Rank, Tom Brown, Mike Leeds, Scot Gresham Lancaster, Norman Locks, and Frank Drake, Arthur C. Clarke, John Cage, Philip Jose Farmer, Robert Kraft, and so many others for their time, energy, and help in this endeavor.

1.

2.

3.

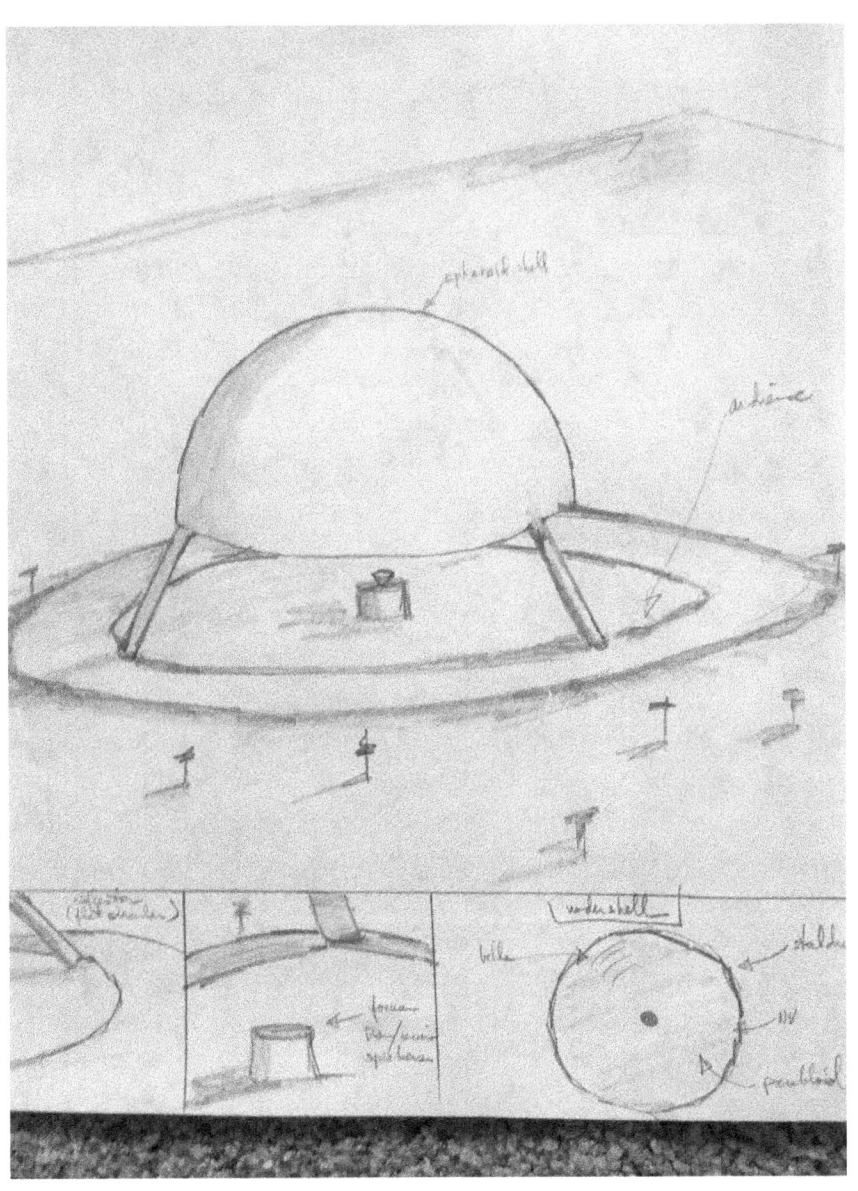

4.

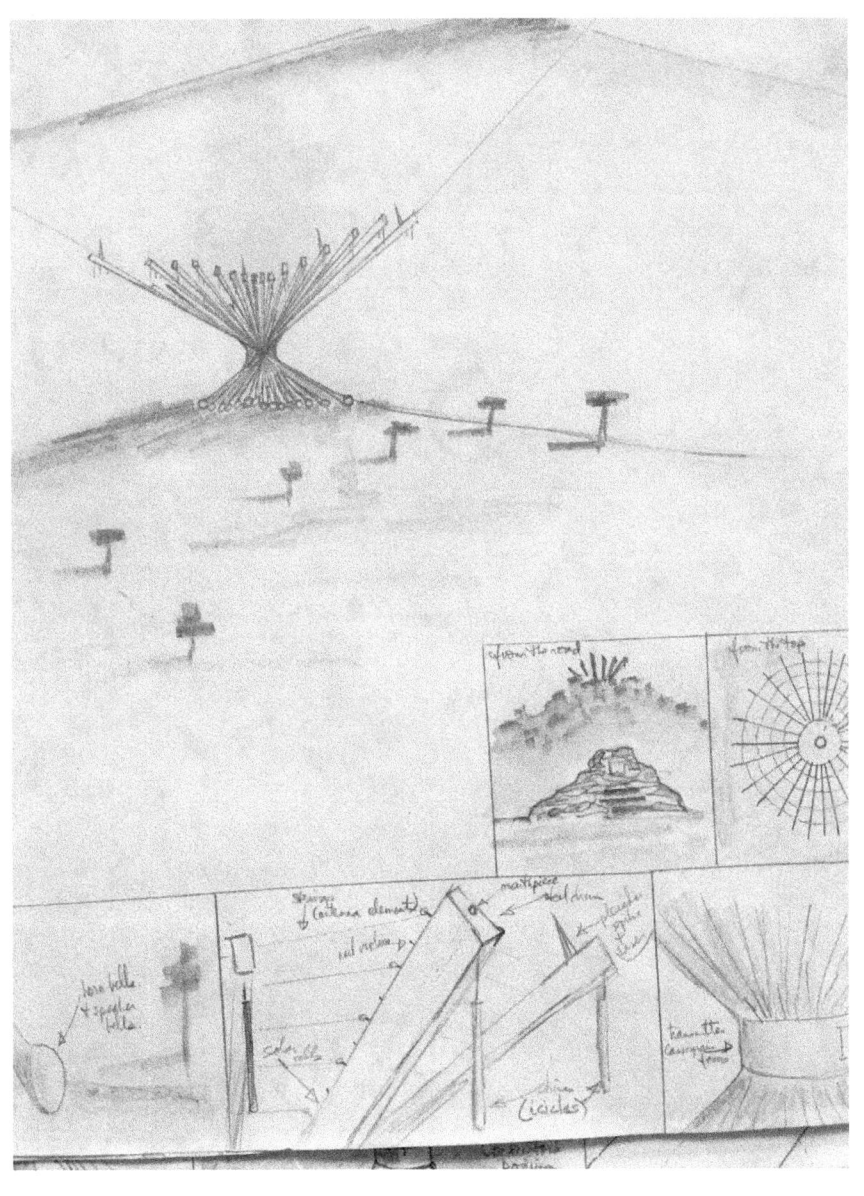

5.

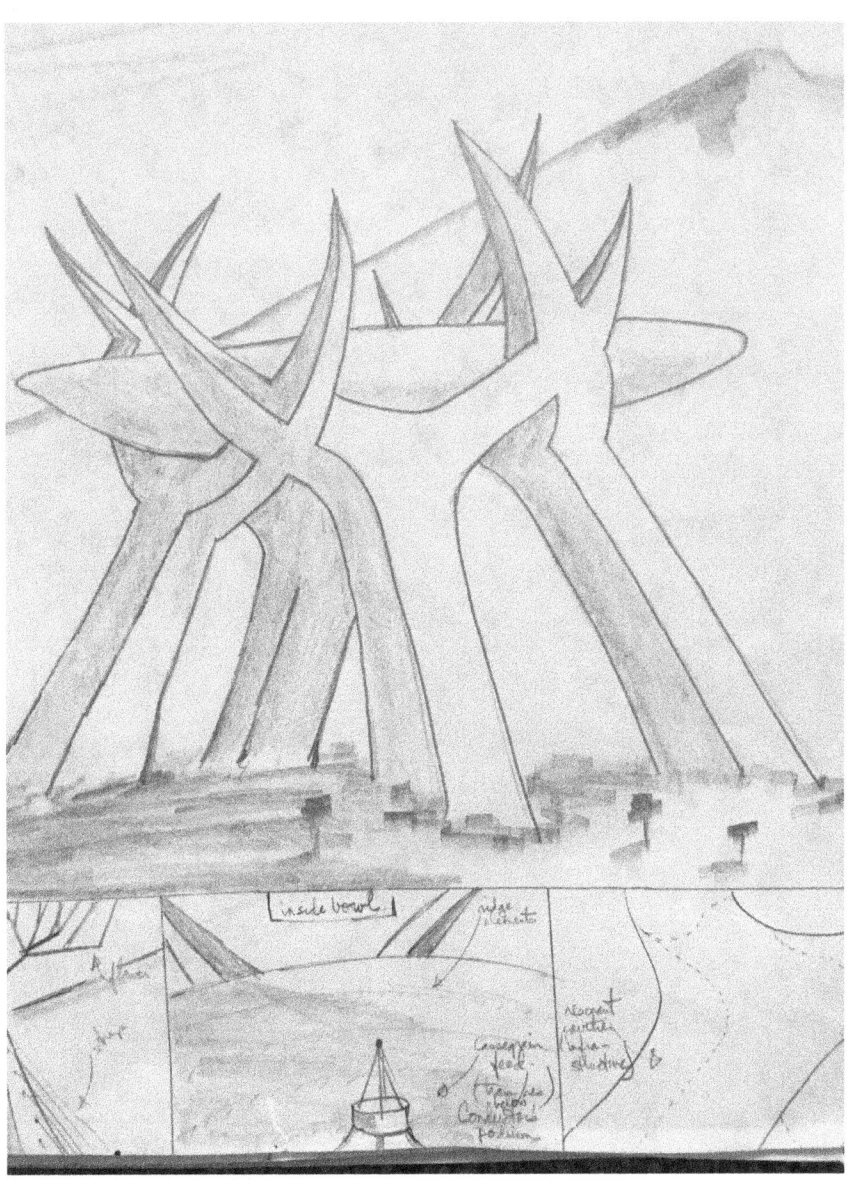

6.

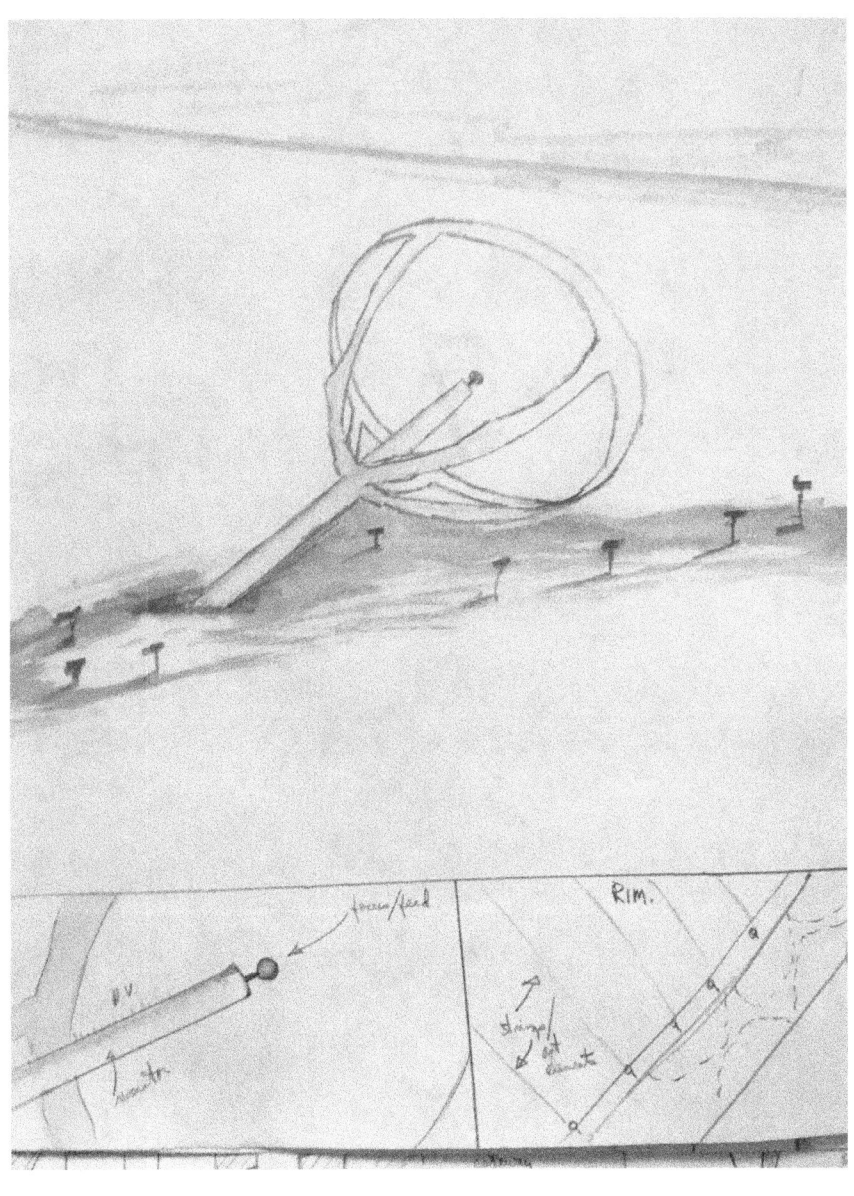

7.

8.

9.

10.

11.

12.

13.

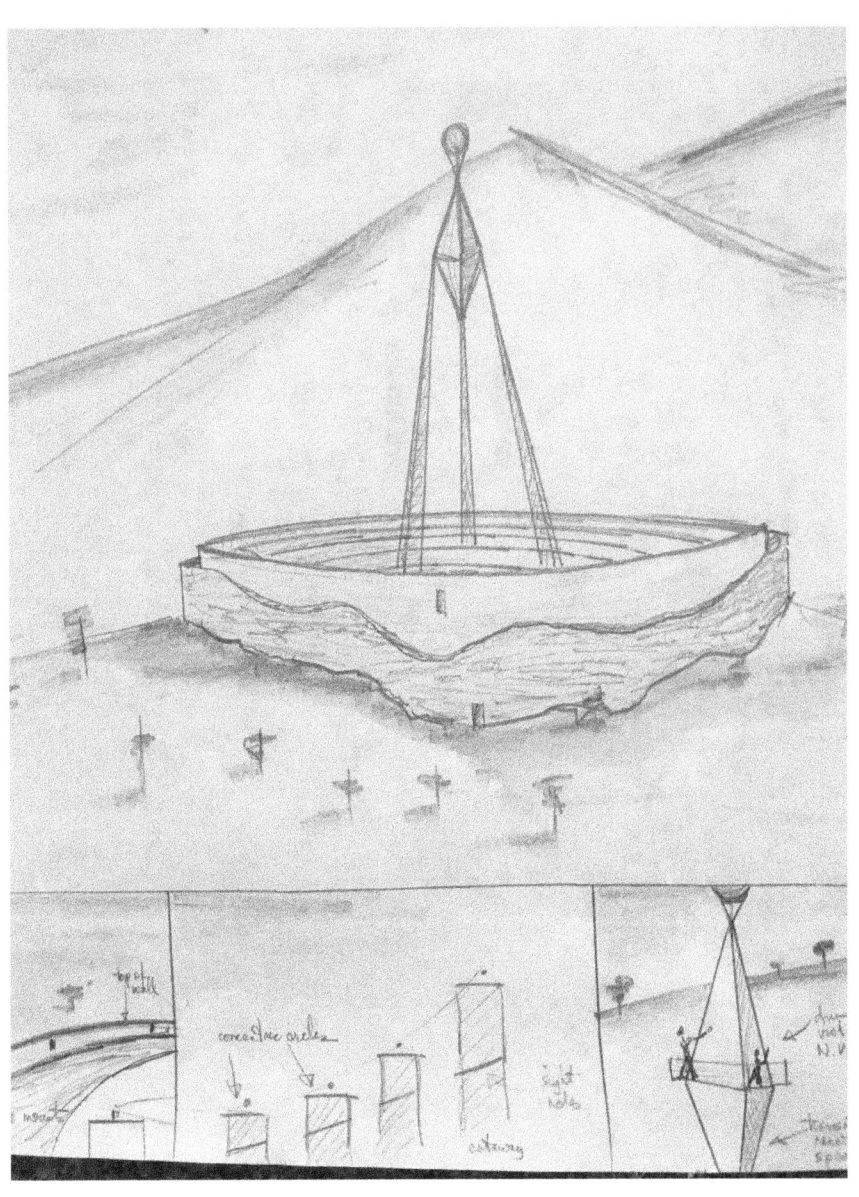

14.

15.

16.

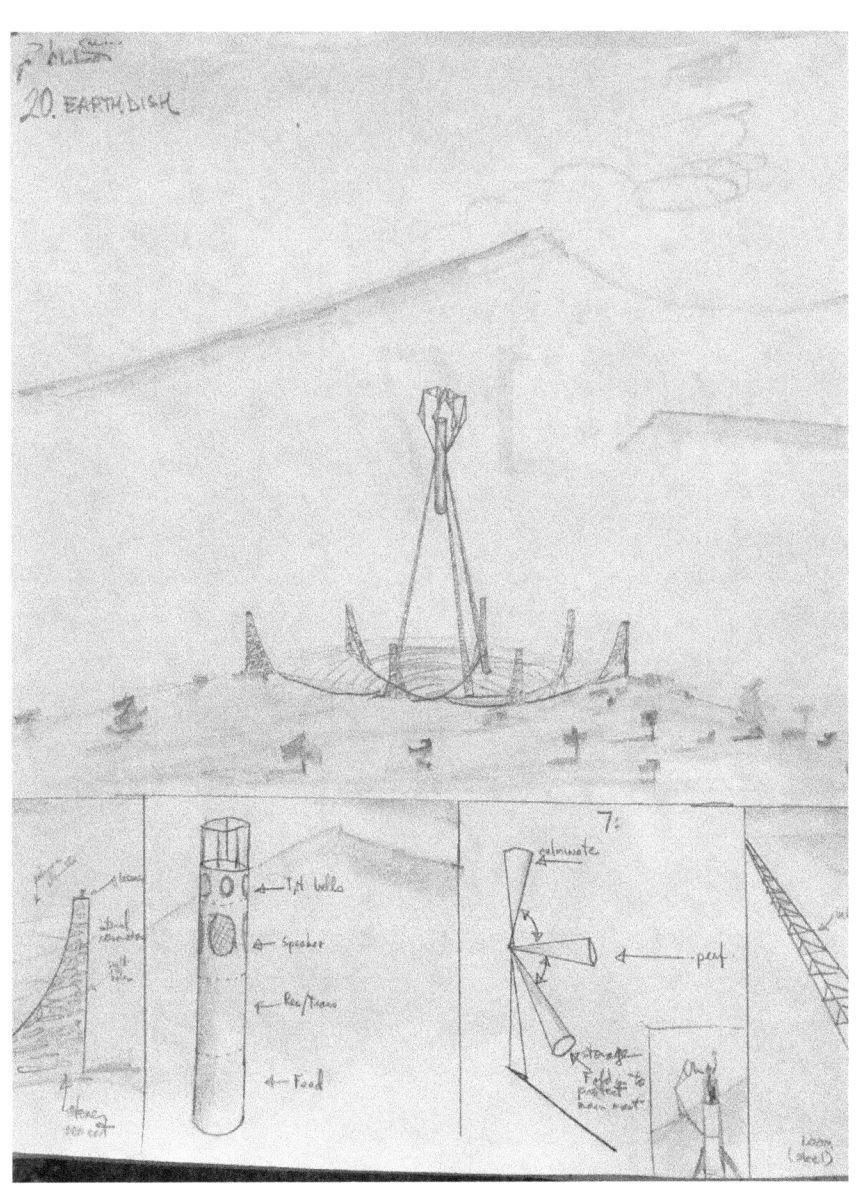

17.

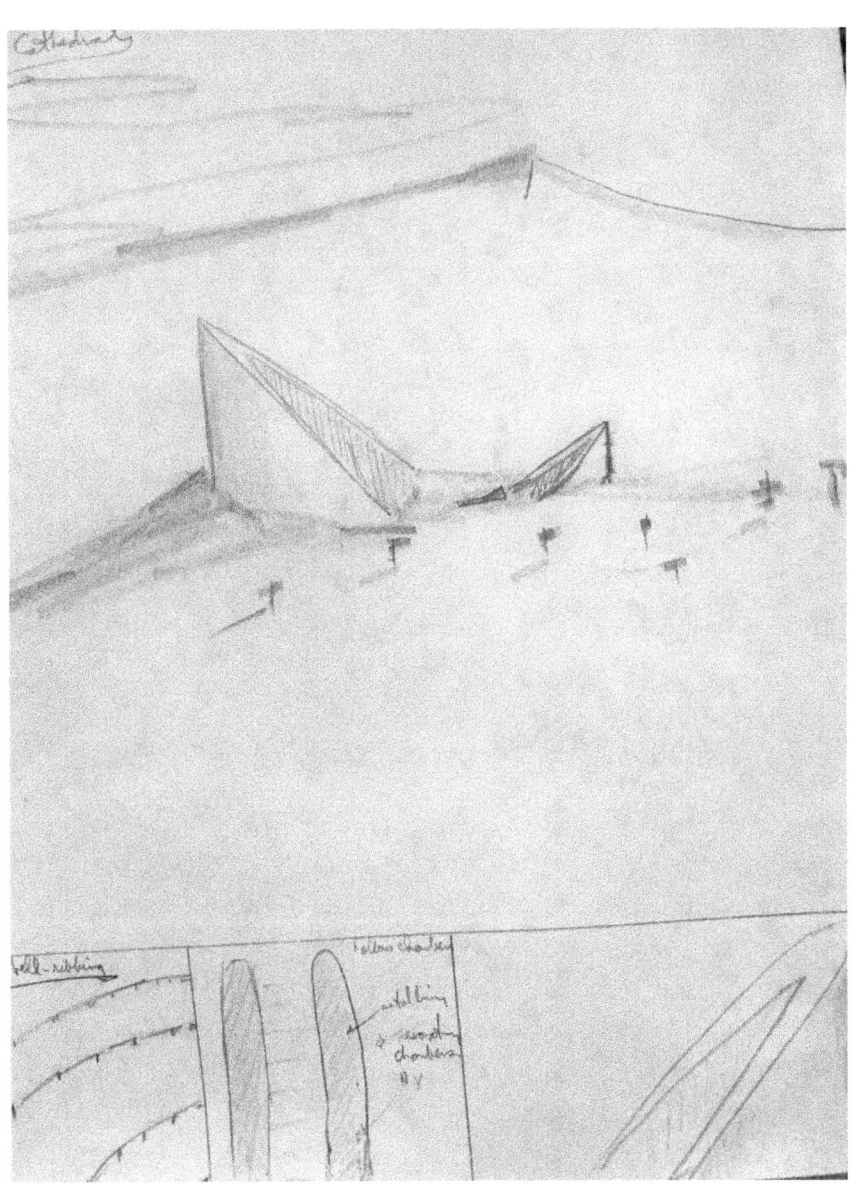

18.

19.

20.

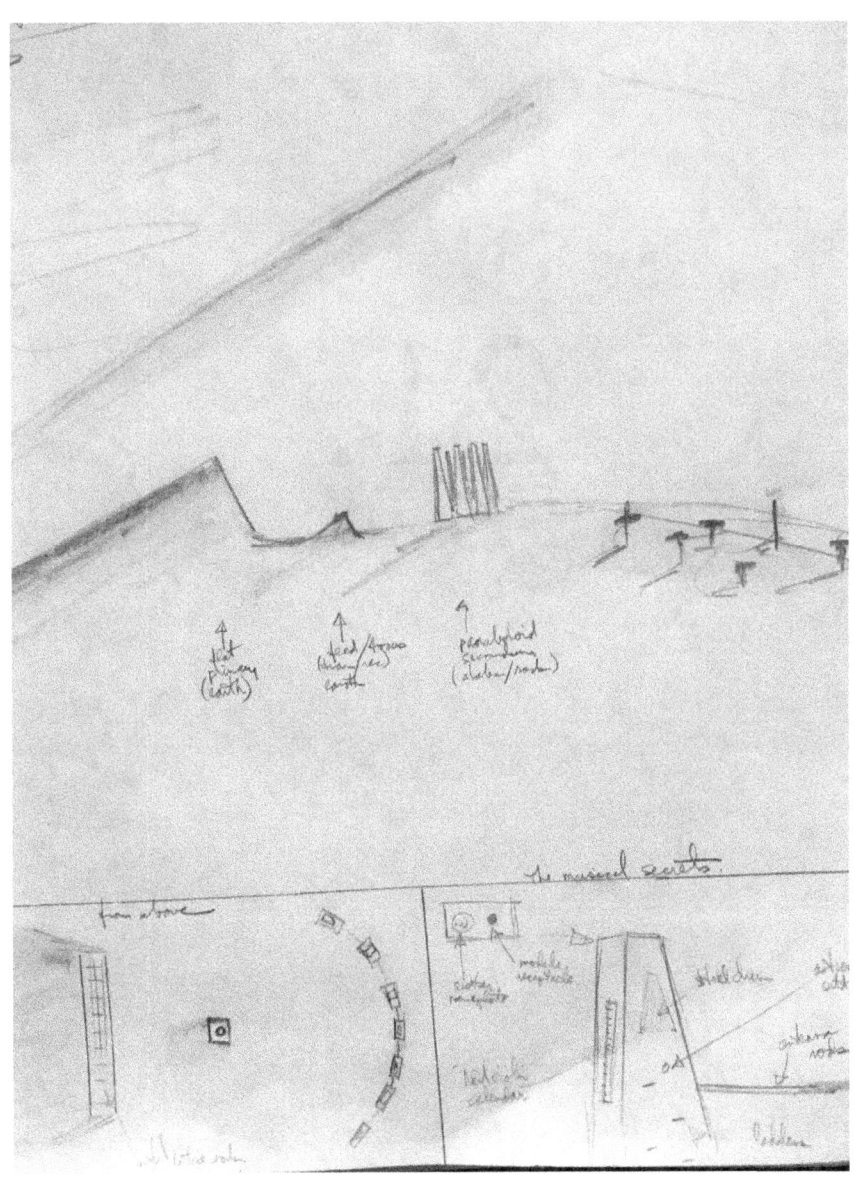

21.

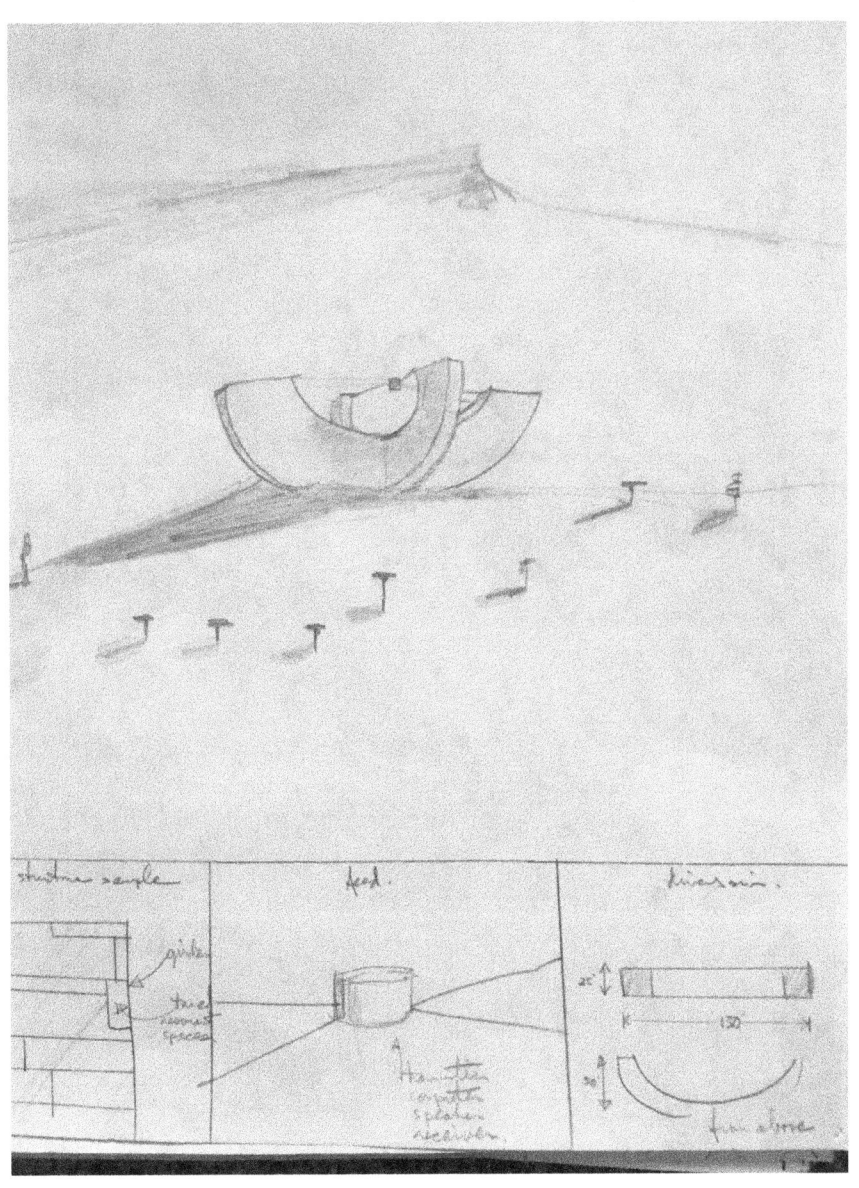

22.

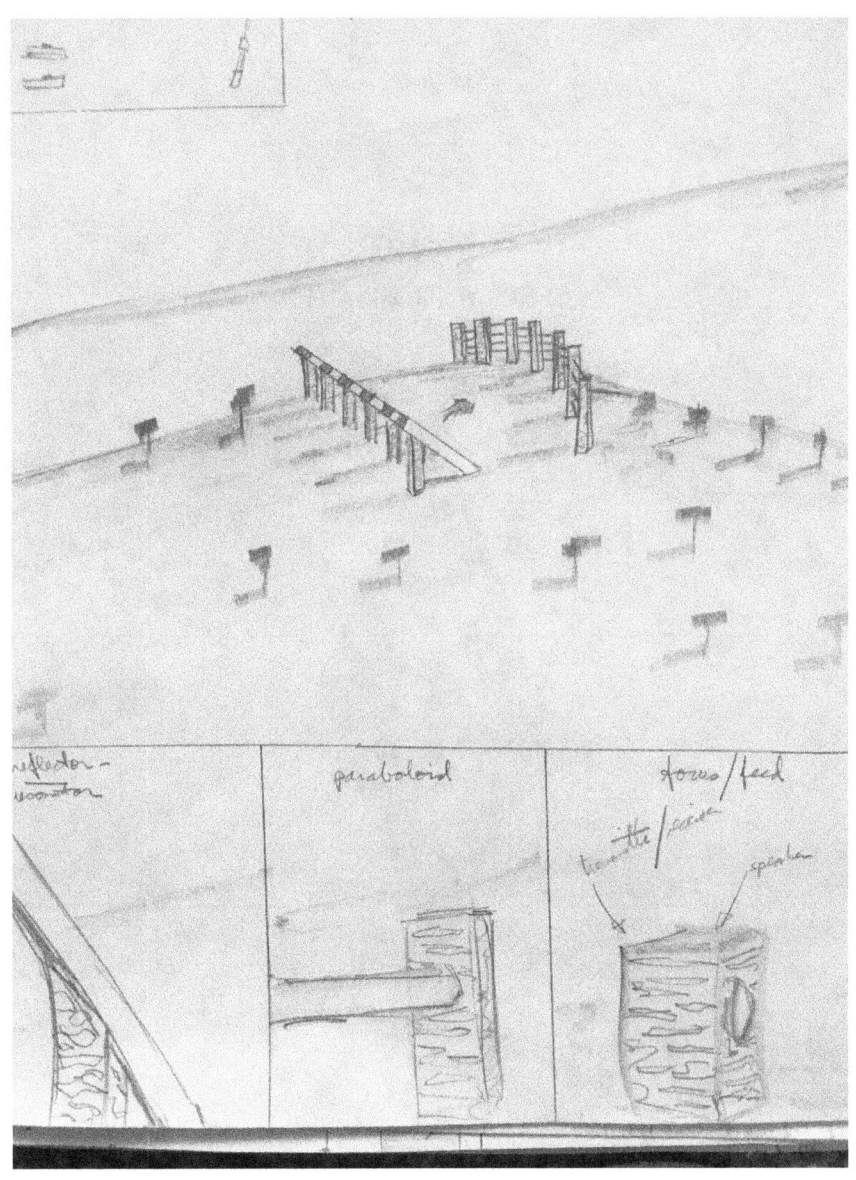

PLEIADES

A GENERAL PROSPECTUS
FOR A LARGE SCALE (105 METER)
ELECTROMAGNETIC-ACOUSTIC
RESEARCH STATION

PREFACE

Any form of energy may be converted into any *other* form of energy. Hydrogen atoms fusing into helium in the sun, for example, convert sub-atomic energy into light, thus making life possible on earth. The photosynthesis of plants (conversion of solar to chemical energy), combustion (chemical energy translated into light and heat), and even digestion (chemical energy creating body warmth) provide further illustration of this universal principle. Called "transduction," this concept has yielded some of the most remarkable advances in the history of humankind. Michael Faraday's discovery of electromagnetic induction in the 19th century resulted in a technological revolution and laid the foundation for electronics. Guglierlmo Marconi's creation of the "wireless," a device transmitting electromagnetic waves without wires (using Hertz's spark coils, themselves transducers) and then translating them into sound (using Edouard Branly's receiver), initiated today's radio and television industries. Thomas Alva Edison utilized the heat produced by electrical current in thin wires to create the first electricity-to-light transducer: the first light bulb.

The *Pleiades Project* has been founded to provide research stations for ongoing study of the mechanics of transduction. The first such station, an as yet unnamed electromagnetic-acoustic telescope, representing the collective wisdom of professional scientists and designers, is an instrument capable of elegant translations and manipulations of energy forms. In its briefest definition, it will allow precise observations of transduction and interactions between electromagnetic and acoustic energies. The station will also provide focus for research in the individual fields of acoustics (and music composition) and radio astronomy.

This integration is most propitious. Electromagnetic and acoustic energies share a number of common phenomena. Both appear as periodic wave forms which may be reflected and refracted and the study of each provides opportunity to better understand the other. The study of waveguide principles, for example, so important to initial advances in radio and radar astronomy, was enhanced by the recognition of similarities in the two forms of energy. The conical shape of acoustic horns has been extensively utilized in designs for directional microwave transmitting antennae. The parabolic reflecting telescopes of astronomy became the model for recent advances in pickup devices for sound waves. Even differences between these forms of energy can be useful. Radio waves, for example, are barely disturbed by the presence of rigid, non-electrically conducting wood and brick walls, while sound waves are severely absorbed. Conversely, radio waves reflect from the thinnest of metal screen doors while acoustic energies barely notice their presence. Hence, applying different forms of waves according to appropriate environmental situations may solve otherwise impossible problems.

An immense amount of important research remains to be accomplished. The *Pleiades Project* Research Station will provide significant opportunity for advancing this goal. In order to realize this extraordinary research and performance tool, an impressive group of professionals have been brought together to collaborate on final designs and oversee construction. The following General Prospectus outlines the functions and ramifications of this unique instrument (the engineering prospectus and Environmental Impact Reports will be available by January 1, 1986). The *Pleiades Project* marks a rare occasion in history when scientists and artists join together to focus on instruments beneficial to both communities while offering *all* gathering places to reflect on man's place in the universe.

I wish to thank the many individuals who have contributed so much to this prospectus and to whom the *Pleiades Project* owes an immense debt: Keith Muscutt (Project Advisor) for his incredible breadth of knowledge, insights and tireless support; Frank Drake (former Director of the Arecibo Radio Telescope) for his continuing advice about the general and specific structural detail; Arthur C. Clarke (novelist and scientist) for his suggestions and commentary; Mark Primack (architectural designer) for his artistic perspectives and perceptive insights; Peter Elsea (electronic music instrumentation) for his expert advice and analytical abilities; John Cage (composer and philosopher) for his support and understanding; Michael Stamp (business consultant) for his tireless help and expertise; Norman Locks (photographer) for his extraordinary color photographs of the site; Philip

José Farmer (novelist) for sticking with the project for so many years; and the following individuals for their advice and contributions: David Rank (astronomer, Lick Observatory); Tom Brown (Attorney); Jack Osborne (engineer, Lick Observatory); Mike Leeds (designer); Stefan Medwadowski (structural engineer); Robert Kraft (Director, Lick Observatory); Paul Horowitz (electrical engineer, Harvard University); Scot Gresham-Lancaster (computer and transducer expert, Mills College); Mary Jane Cope (performer, UCSC); Julia Garnett (public relations consultant); Linda Marcetti (media expert); Robert Jastrow (Dartmouth University, astronomer); Tim Cope (artist); the University of California at Santa Cruz Arts Division and the Executive Committee of Porter College (UCSC) for funding the publication of this General Prospectus.

-David Cope

Project Director, PLEIADES
Professor, Porter College
  University of California at
  Santa Cruz

Pleiades (pronounced plee'ya-deez) is a spectacular cluster of stars in the constellation TAURUS. For millennia this "bejeweled" configuration of stars (often called the "seven sisters") has inspired scientists and artists alike. Greek astronomers used the heliacal rising of the Pleiades to mark the inauguration of the navigational season on the Mediterranean while Aratus (third century BC) bestowed poetic names on each of the individual stars. The Pleiades also played a prominent role in ancient cosmology of the Americas with many monuments, including the Nazca desert intaglios (Peru) and the design of the city of Teotihuacan (Mexico), believed to have been oriented to the rising of this star cluster. Hindu, Buddhist, Hebrew and New Testament scriptures frequently mention the cluster as do many poets including Poe, Blake and Tennyson. This rich history of combining astronomical observation with art, architecture, design and poetry continues in the *Pleiades Project*, fusing the empirical and the intuitive.

TABLE OF CONTENTS

CHAPTER 1
OVERVIEW

INTRODUCTION

The universe abounds with periodic vibrations of energy radiating at measurable speeds and frequencies. One perceives sound (acoustic energy) when these excitations occur between 1,000 to 1,200 ft. per second and at oscillations of 20 to 20,000 cycles per second (Hertz). Light waves (electromagnetic energy) travel at 982,080,000 ft. per second within a 430,000 to 750,000 GigaHertz (billion cycles per second) frequency range. For centuries astronomers have concentrated on this latter visible spectrum. Recently, significant advances have taken place in the radio frequency bands: electromagnetic energies from 50 MegaHertz (million cycles per second) to 100 GigaHertz and above. Telescopes, sensitive to modulations in signal amplitude of radio waves, have revealed new information on known sources and discoveries of hitherto unknown objects emitting only radiowaves (invisible to optical astronomers). Since electromagnetic and acoustic waveforms radiate in like manner and respond similarly to all forms of propagation (diffraction and focusing), comparing results from research in each energy form reveals yet further information and insights.

The *Pleiades Project* presents a "state of the art" instrument for the study of stellar radio sources and terrestrial acoustic resonances. Designed as a photon/acoustic transducer, it will translate received radio energy into sound and vice versa. The centerpiece is a 105 meter wide stationary antenna following the highly successful Arecibo (Puerto Rico) design. This "dish" may be composed of any material conducive to reflecting acoustic and/or electromagnetic energies. This convertibility is highly attractive to astronomers who constantly improve the deviation specifications of surface materials but find it prohibitive to continuously seek funding for completely new instruments. Acousticians and composers will also find this flexibility promising as they may introduce a wide variety of different sound producing surfaces. The plan proposed here includes sophisticated metal plates which serve simultaneously as effective reflectors and elegant acoustic instruments.

This electromagnetic-acoustic telescope was conceived by *Pleiades Project* Director David Cope as a focus for interdisciplinary research. Although principally known as a composer, he has built many musical and astronomical instruments and written extensively on the work of other experimentalists in books such as *New Directions in Music* and *New Music Composition*. A group of distinguished consultants has and will continue to work closely with the Project Director ensuring that every step and detail is handled with the expertise of professionals. As well, teams of radio astronomers, engineers, architects, electronic consultants, acousticians, etc. will be assembled to realize final designs and oversee construction. Computer engineers will install and monitor the microprocessors necessary to operate the intricate acoustic and astronomical equipment. A group of prominent business administrators will provide their experience and acumen for the on-going functioning of the *Pleiades Project* (under the auspices of the *Taucet* *Corporation*).

A. General Description of the Instrument

This research station for experimentation and creative work will explore the shared phenomena of audio and radio frequencies. Music (aesthetically and empirically organized sound) occurs as acoustic vibrations. These complex waves are analogous to radio waves. Creating in and studying the two energies simultaneously has previously been accomplished by combining light and sound in "multimedia" composition. This electromagnetic-acoustic telescope will afford composers and astronomers the opportunity of exploring the new dimension of sound/radio wave spectra.

Since signals from the stars are very weak, radio telescopes require large gathering surfaces. Hence, in its initial construction, the instrument will take the form of a 105 meter diameter spherical reflector consisting of more than 6,000 specially designed instrument-plates. These will act as bilateral transducers between the signals of far away stars and galaxies and local acoustic vibrations. The telescope will be housed in a large radio-translucent geodesic dome with built-in sensors made from a remarkable new substance called Kynar, a transducer membrane capable of functioning as both microphones and speakers. Scientists and artists may study and manipulate the translation of one form of energy to the other, providing an extraordinary palette of potential uses. A composer, for example, might "play" the extraordinary instrument-plates of the reflector while simultaneously broadcasting corresponding radio images toward the moon and incorporate the returned signal as feedback into the telescope dish. An astronomer might listen to the wealth of "radio sky" signals studying active point sources for interstellar signs of organic molecules. Even when not involved in specific projects, the instrument can sense the sky and sympathetically reverberate with the "sounds of the stars" providing data for continuing research of large scale resonances and aural analysis of information patterns in collected radio data. Since human aural perceptions vary significantly from visual imaging, even surpassing the latter in many circumstances, a significant potential exists for recognizing special patterns in the sonic translations of electromagnetic energies.

B. Historical Background

1. Music

Exploration of large-scale musical resonances began as early as the 14th century with works like Thomas Morely's *Canon for 16 Choirs*, an enormous antiphonal work for the dedication of the Oxford Cathedral. Giovanni Gabrieli's eight and twelve part canzoni (late 16th century) represented further acoustic experiments. These continued through works by Mozart (*Don Giovanni*), and the *Requiems* of both Berlioz and Faure. In more recent years, Charles Ives (*The Unanswered Question* and the *Universal Symphony*), Henry Brant (*Voyage Four*), and Gordon Mumma (*Hornpipe*) have explored the movement of sound (timbre and spatial modulation) as well as sounds produced in large spaces. The Institute for Research and Coordination of Acoustics/Music located in the Georges Pompidou Center in Paris includes *Espace de Projection*, a performance hall with variable acoustic properties by use of computer controlled shutters and "flats". The Stanford University Center for Computer Research in Music Acoustics has produced outstanding results through modeling techniques in various real-time computer programs. Likewise, the history of automated music is rich and varied with music boxes, nickelodeons, etc. paving the way for composers such as Luigi Russolo (Italian Futurist of the 1920s and 30s), Alexander Mosolov (notably his ballet: *Steel*), Conlon Nancarrow (*Studies for player piano*) and most recently the "League of Automated Composers" (principally John Bishoff, Don Day, Jim Horton and Tim Perkins).

Valid experimentation with large scale architectural spaces, however, depends on the predictable qualities of the sound source and the controlled variability of the resonant space. Studies in both Doppler and reverberation effects have suggested that without careful control of the pitch source, resultant sonic events are unpredictable and virtually unresearchable. To date, truly large scale instruments have been unavailable in any but the most rudimentary forms.

2. Radio Astronomy

Radio telescopes inherently demand large gathering and reflecting space to be useful. Since Jansky's discovery of residual noise signals from the Milky Way in the 1930's and Grote Reber's construction of a 9.14 meter (30') diameter steerable paraboloid, large aperture instruments for radio astronomy have developed into an indispensable requirement for research. Current engineering precision and the availability of accurate computerized templates have increased potential and sophistication. The multitude of possible frequencies combined with the infinite choice of objects for study has created immense need for new and accessible instruments.

3. Transducers

The concept of relating stellar phenomena and music has previously been proposed by Pythagoras and Plato (whose *Republic* relates the motion of heavenly spheres to the tones of a musical scale [*musica mundana*]). The 17th century mathematician-astronomer Johannes Kepler in his "Music of the Spheres" (*Harmonice Mundi*) suggested that planetary motions, balances and interrelationships corresponded to similar correlates in harmony. He further extrapolated music based on these observations.

2

A significant body of experimental research has also been accomplished in the "media" compositions of notable 20th century composers and visual artists including *Prometheus - The Poem of Fire* (Alexander Scriabin; 1910) and *Poème électronique* (Edgard Varèse, composer, in association with architect Le Corbusier; 1958). The *Pavilion* of Expo 70 in Japan created an extraordinary "laboratory" for interdisciplinary work in art and technology and led to collaborations between visual artists, composers and engineers. Other experiments by Nam June Paik, James Searight, Allan Kaprow and Larry Austin continue to explore interrelationships between media.

The scientific and musical investigation of both acoustic and radio resonance seems natural and propitious. While laboratories built by Experiments in Arts and Technology, Bell Labs and PARC provide valuable resource for research and nourishing cross-disciplinary experiments, none approach the potential for large-scale study provided by the *Pleiades Project* Research Station.

C. Operations of the Instrument

The instrument may be operated in four possible modes:

1) As a radio telescope capable of precise observations up to 60 GigaHertz frequencies. With its 105 meter diameter primary reflector, observations of numerous important active radio sources could take place at these microwave frequencies. It is also anticipated that recent discoveries of organic molecules in interstellar space will attract many research projects.

2) As a powerful radar transmitter capable of reflecting signals off nearby planets such as Mars and Venus and interplanetary objects such as the moon, asteroids and passing comets. Such capabilities will enhance research in topographical and geological mapping of local astronomical bodies.

3) As an acoustic instrument capable of an immense array of pitches, rhythms and timbres in large scale resonance. In the current proposal, each sensitive instrument-plate in the surface of the large reflector may be activated by electromagnets connected to a large digital synthesizer. Likewise, the dome covering the reflector contains sensor membranes (Kynar) which act as microphones (receivers), speakers (transmitters) or both. Resonance may be altered by thousands of variable (radio translucent) panels in the geodesic ceiling which is controlled by the computer system.

4) As an integration of the above functions capable of producing subtle interactions between electromagnetic and acoustic energies. Since the instrument may be described as a large transducer surrounded by terrestrial and interstellar spaces, any manifestation of one waveform may be observed, studied, paralleled, reverberated and/or reflected in the other. The number of possible combinations is limited only by the imagination of researchers.

Aside from astronomy and acoustics, this instrument may be used for ancillary research fields including:

> Computer Sound Synthesis
> Acoustic Holography
> Ultrasonic and Subsonic Acoustics
> Wave Properties
> Electromagnetics
> Pyroelectricity
> Piezoelectricity
> Search for Extraterrestrial Intelligence (SETI)
> Musical Instrument Design
> Large Scale Geodesics
> Artificial Intelligence
> Cybernetics
> Architectural Acoustics

D. Summary of Costs and Contingencies

Due to the nature of the instrument's design and the complementary shape of the Walker Pass site, the estimated cost of the telescope is remarkably low; less than 10% of the projected costs of more conventional instruments of comparable capabilities.

The following estimates represent the primary costs involved in construction (in K dollars). Figures are derived from estimates of northern California contractors and prospecti of other radio telescope projects (based on the 1984 dollar).

```
Surface and Approach Road Preparation*..............50
Concrete Foundations in Bedrock.....................20
Plate Milling and Machining**.....................900
Foundation Geodesic and Resonators................150
Electronics.......................................150
Clocks and Timing Oscillators......................60
Bearings and Gears.................................12
Feed..............................................100
Drives.............................................50
Computers.........................................250
Instrument Cabin...................................50
Living Quarters....................................70
Utilities..........................................20
Design and Construction Management................100
Radome............................................250
Kynar Membranes....................................40
```

SUB-TOTAL	2,272
CONTINGENCY @ 10%	227.2
TOTAL FUNDING NEEDS	2,499.2

Since the test section (discussed under #2.G) will eventually be utilized as a part of the final instrument, its costs have been included here (anticipated at $50,000 depending on the availability of rental electronics and computing systems for the initial tests).

E. Funding

The *Tauceti Corporation* will seek tax deductible funding principally from private individuals and foundations as well as local, state and federal arts and science agencies such as the California Arts Council, University of California, the National Endowment for the Arts and the National Science Foundation. As well, grants of materials will be sought from companies manufacturing electronic, steel, aluminum and computer products. Much of the unskilled manual labor may be provided by volunteers.

Fund raising will continue beyond construction in order to create an endowment for the on-going operations of the instrument and to allow researchers maximum access at the least possible expense.

The instrument will function on a project-to-project basis. A select group of staff members capable of operating either the telescope and/or the acoustic instruments will be available at 60 days notice on a daily rate basis (through the *Tauceti Corporation*). Actual costs may vary depending on the type of project and the number of staff necessary.

*Includes the construction of the on-site road as well as the resurfacing of the dirt road from Highway 178.

**Based on fabrication by outside contractor (in-house construction could reduce funding needs by 60%).

CHAPTER 2
ASTRONOMICAL CONSIDERATIONS – RADIO TELESCOPE

INTRODUCTION

The National Science Foundation has identified the construction of large telescopes of this kind as one of the highest priorities in science. The available
frequencies combined with the number of celestial objects for study insure the
ongoing need to continue intensive radio sky research. Currently, the high cost
of building and operating radio telescopes has impeded a number of important mapping projects, astrogeology, observations of unique stellar events, and the Search
for Extraterrestrial Intelligence. This instrument offers a "public access" telescope capable of highly professional results.

A. The Design

1. The Basic Structure

The construction of this instrument responds to the need for development of large
scale radio telescopes following rigorously specified forms. While its size and
weight prevents physical motion, the development of a spherical reflector (Arecibo
type) and a movable Gregorian secondary feed permits a 60 degree declination sweep
(30 degrees in any direction from the central axis) as well as a 2.5 - 4 hour
tracking time. Thus, by concentrating on slightly smaller portions of the
reflecting surface and by pointing the secondary reflectors rather than the primary, the telescope responds in a similar manner to fully steerable types.

The following section plan illustrates the major features of the instrument:

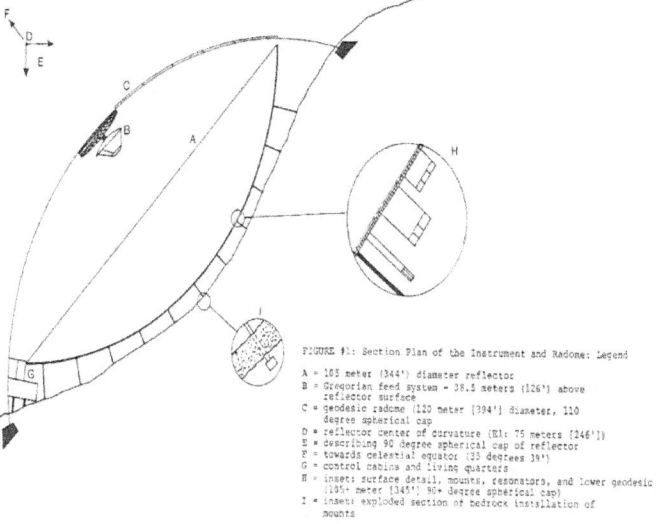

FIGURE #1: Section Plan of the Instrument and Radome: Legend

A = 105 meter (344') diameter reflector
B = Gregorian feed system - 38.5 meters (126') above
    reflector surface
C = geodesic radome (120 meter [394'] diameter, 110
    degree spherical cap
D = reflector center of curvature (R1: 75 meters [246'])
E = describing 90 degree spherical cap of reflector
F = towards celestial equator (35 degrees 39')
G = control cabins and living quarters
H = inset: surface detail, mounts, resonators, and lower geodesic
    (105+ meter [345'] 90+ degree spherical cap)
I = inset: exploded section of bedrock installation of
    mounts

Construction difficulties will be minimized by the nature of the site (itself a rough spherical cup) and the geodesic support of the feed system.

2. The Reflector Design

The geodesic sub-structure of the reflector presents a "cradle" for the installation of virtually any kind of surface. Precision cut and aluminized styrofoam could, for example, be quickly installed. Other yet to be discovered materials, providing even greater accuracy, could replace existing plates in a few weeks negating the need to procure new sites and build expensive new structures. Initially, to reduce cost and installation time, a small (concentric) circle of of plates could be formed for testing. If successful, the remainder of the dish could be completed as desired. Almost any material, barring prohibitive weight problems, could be substituted (see suggestions in #3.A.1). The following describes a resonant metal plating, chosen for its inherent acoustic and reflecting properties.

The first proposal of a surface for this Research Station follows the shape of the 105 meter (345') diameter 90 degree spherical cap (75 meter [246'] radius of curvature) constructed of appropriately curved (1 degree per 1.31 meters [4.294']) square and rectangular plates of 3mm (1/8") to 1.5mm (1/16") thick galvanized steel. Experiments during the construction of the test section (see #2.G) will determine whether machining or mold forming techniques provide the greatest accuracy.

Each plate will rest on 4 to 8 adjustable (radial screw) mounts projecting from a like formed geodesic which acts as superstructure for the reflector. This support system, a multi-frequency icosahedron (90 degree cap of a sphere with a 75.3 meter [247'] radius of curvature), will fit in the propitiously shaped curve of the southern face of Walker Pass in southern California and be supported by footings rooted in bedrock below the decomposed granite of the site. Thus, changes in the terrestrial surface conditions won't affect the reflector plates due to the "shock absorber" effect of the intervening geodesic.

Data from a number of radio telescope design proposals and standard temperature coefficients for galvanized steel plates of similar thickness to those in this instrument indicate surface degradations, even under the most drastic conditions ($\Delta t = 12^\circ$/hour), should not exceed .5mm. Slight separation between plates, a fraction of the smallest wavelengths predicted for the telescope, will allow free expansion and contraction without significant deviation from true curvature. Plates will be installed at a temperature just beyond that predicted for greatest thermal expansion. At this stage, they will be designed to touch uniformly so that they will pull apart slightly as the temperature varies. This also allows the plates to vibrate during acoustic performance. As well, the plates will only be attached at nodal points allowing resonating movement for acoustic performance and permitting further self-adjustment during extreme thermal stress. The radial screws provide opportunity for manual adjustment. Thus, most expansion and contraction will not create significant surface variations. Engineering experimentation with the new technology of thermally invariant structures should increase tolerances and even reduce cost.

1.5 mm (1/16") steel weighs roughly 1.152 pounds per square ft. Hence the plates will weigh approximately 124K pounds (62 tons) spread over the steel geodesic below and combined with its weight projects into the hillside below. During the construction of the test section, a number of steel, aluminum, nickel and/or magnesium alloys will be tested both for their difference in weight and for their inherent musical qualities. Upper plates (more vertical and hence more difficult to mount and control) may be small compared to the lower (more horizontal) larger plates. They may also be composed of lighter materials (see #3.A.1). Exact number and size of plates will depend on experiments carried out in the test section (see #2.G) and subject to engineering, acoustic and astronomical (surface accuracy control) considerations. The plates will sit 9.6 cm (4") above the geodesic support system which in turn rests never less than 2 ft. above ground level (to allow access for repairs, tuning and adjustments).

3. Surface Qualifications

In the initial surfacing, the machined or molded contour plates of the reflector should approach .3125mm (.0123") deviation from perfect spheroid shape. Using

$$\lambda = D \times 16$$

where λ = wavelength and D = deviation, and translating the resultant 5mm to frequency using

$$F = 300/\lambda$$

where F = frequency in MHz, results in a maximum frequency limit of 60,000 MHz (60 GHz).

Other types of deviations can result from 1) errors resulting from the production of the single plates; 2) adjustment errors of the plates; and 3) elastic deformation due to the temperature changes mentioned previously. It is anticipated that none of these deviations will exceed the 5mm previously discussed.

With continued analysis and testing during fabrication and installation, a lower tolerance may be achieved allowing higher frequency observations. Manufacturers such as the Electronic Space System Corporation have quoted deviations as low as .06mm (1/5th that expected in this instrument) using various materials. The Boeing Corporation, the Lockheed Missile and Space Company and the Philco-Ford WDL Division also predict low deviations.

4. The Feed

The feed will consist of a (approximate) 9.5 meter (31') diameter Gregorian secondary reflector located above the paraxial focus and a (approximate) 1.52 meter (5') diameter tertiary reflector located approximately 6 meters (20') below the secondary. The tertiary delivers reflected waves directly to a feed horn located at the vertex (Cassegrain model) of the secondary eliminating extensive wave guides. This aberration-correcting Gregorian feed system will be situated 38.4 meters (126') above the primary on a curved grid designed as a parallel section of a convex spherical cap with identical center of curvature as the main reflector. This beam moves around a motorized pivot mechanism attached to a platform allowing a 30 degree swing in any direction for 60 degree declination movement and as much as four hour tracking time (with acceptable vignetting) of radio sources (see #2.A.6). Phased line and/or point source feeds (horn, Yagi or log periodic antennas) may be added for specialized studies at lower frequencies but will not be a part of the initial complement of feed equipment.

The 32 meter (105') diameter aluminum grid platform will be secured to the protective geodesic radome providing stabilized support. The unit imitates the basic reticulated form of the radome thus creating as little increase in reflector shadow as possible. Both upper and lower geodesics will undergo ASTRA (Advanced Structural Analyzer) computer program analysis to determine full static stress, deflection due to loads and thermal effects as well as natural vibration nodes and frequency buckling potential and dynamic response to harmonic focusing. A small pathway along the upper section of the radome provides access to the feed. A set of removable grids creates direct entry into the feed grid area.

A "pendulum-swing" connector, located at the lowermost extremity of the platform, will allow installation of precision cutters for various types of surface materials (e.g. styrofoam proven effective at the Danby project at Cornell). Even though an entire set of plates could not be cut simultaneously (due to the angle of the reflector), project consultants expect only eight or less separate sessions to provide adequate area to re-surface the entire cradle of the support geodesic.

5. Wavelength/Resolution/Sensitivity

The following is a list of non-shared radio frequency bands allocated to Radio Astronomy by the FCC and the International Frequency Registration Board of the International Telecommunications Union.

Frequency band	
73 - 74.6 MHz	
79.75 - 80.25 MHz	
150.05 - 153 MHz	
322 -329 MHz	(Deuterium line)
404 - 410 MHz	
606 - 614 MHz	
1400 - 1427 MHz	(Hydrogen line)
1664.4 - 1668.4 MHz	(OH line)
1660 -1690 MHz	
2690 -2700 MHz	
3165 - 3195 MHz	

```
4800 - 4810 MHz
4990 - 5000 MHz
5800 -5815 MHz
8680 - 8700 MHz
10.68 - 10.7 GHz
15.35 - 15.4 GHz
19.3 - 19.4 GHz
31.3 - 31.5 GHz
33.0 - 34.0 GHz
36.5 - 37.5 GHz
```

Choice of wavelength depends on allowable size of beamwidth which in turn relates to the diameter of reflecting surface:

$$\theta = \frac{60}{D}$$

where θ = beamwidth, D = diameter, and λ = wavelength.

With D at 105 meters, and wavelength set at .005 meters, then beamwidth = .00286 degrees or 10.3 seconds of arc, far better than any other single telescope and surpassed by only 2 or 3 substantially more expensive "systems". This resolution equals that found in good optical systems and thus significantly increases the "power" of radio astronomical observations. Lower frequencies allow for broader beamwidths, less severe surface tolerances and less interference from galactic background noise. Higher frequencies provide narrower resolution for detailed observations of a wide variety of point sources such as quasars and emission nebula as well as contributing important information on complex interstellar molecules. The anticipated gain should approach 50 dB.

The following chart lists many of the world's major radio telescopes by size (larger gathering space preferred) and wavelength (shorter or microwave potential most desirable):

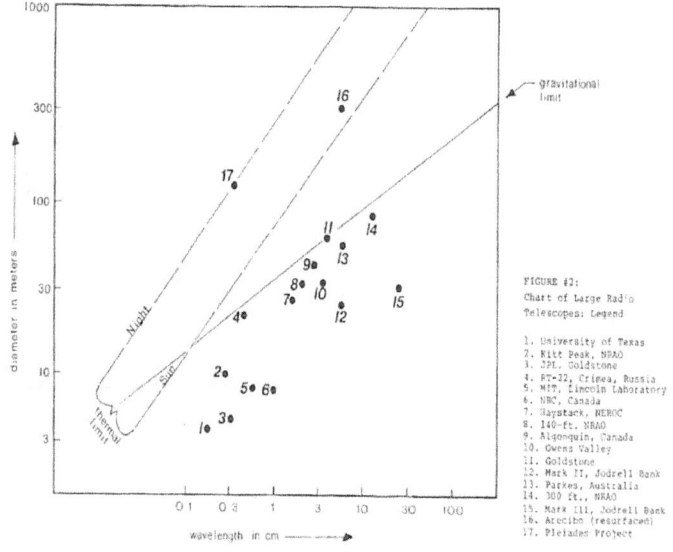

FIGURE #2:
Chart of Large Radio Telescopes: Legend

1. University of Texas
2. Kitt Peak, NRAO
3. JPL, Goldstone
4. RT-22, Crimea, Russia
5. MIT, Lincoln Laboratory
6. NRC, Canada
7. Haystack, NEROC
8. 140-ft. NRAO
9. Algonquin, Canada
10. Owens Valley
11. Goldstone
12. Mark II, Jodrell Bank
13. Parkes, Australia
14. 300 ft., NRAO
15. Mark III, Jodrell Bank
16. Arecibo (resurfaced)
17. Pleiades Project

The gravitational limit shown describes the maximum practical weight for movable reflectors where size is directly proportional to stress and system degradation. The thermal limits describe the practical limits of temperature changes influencing the accuracy of the reflector surface. Larger telescopes have more surface area and thus lose effectiveness quite measurably in the higher frequency ranges. The radome will increase protection against such deviations and permit observations at even higher frequencies than those previously indicated if the reflector surface qualifications permit. Thus, by virtue of its size and frequency range, this telescope ranks as a world-class scientific instrument capable of unique and significant contributions.

6. Direction and Focus

While the main antenna remains stationary, the movement of the Gregorian feed will permit steering anywhere within 30 degrees of the central axis. Utilizing the natural rotation of the earth for right ascension motion and setting the reflector axis in precisely the southerly direction towards the celestial equator (35 degrees, 39 minutes, 40 seconds south of the zenith) allows study of an expansive area for research (60 degrees of declination or 60% of the area covered by a fully steerable telescope). A significant number of active galactic (including much of its nucleus) and extragalactic radio sources fall within this window including all of the ecliptic plane for lunar, solar (especially chromosphere studies) and planetary observations. The Galilean satellites of Jupiter fall in this area and could be easily observed along with the thermal emissions from Saturn's rings. As well, important objects such as W49, W51 (The Orion complex; not visible from the Arecibo telescope), and the active radiogalaxy M87 occur in this region. *Pleiades* (between +23 and +24 degrees declination) appears in the upper range of the telescope coverage.

The neighboring terrain to the south (see #4.A) should pose no obstacles to this angle with the closest peak obscuring less than 16 degrees of the sky allowing 10 degree separation from the southernmost beam swing.

7. The Radome and Support Geodesic

A geodesic "space frame" radome will cover the entire telescope structure for environmental protection (against wind, rain, snow and debris), security, acoustic resonance, and as support for the Gregorian feed system. The requirements for maximum electromagnetic performance (i.e. minimal dB loss or distortion of incoming signals) dictate that the beam cross section of the reticulated spherical structure be as compact as possible and that the length of the beams be as long as possible compared to the operating wavelength. As well, the thickness of the intervening membrane should be very small compared to the operating wavelength.

A standard multi-frequency (110 degree spherical cap) icosahedron constructed from aluminum will provide the basic superstructure of the radome conforming as closely as possible to the above mentioned criteria. Determination of the frequency of the icosahedron follows the formula:

$$a = \frac{A}{F^2 \times S}$$

where A = area of sphere $(4r)$ F = frequency squared (total number of subdivisions) S = number of sides of polyhedron chosen (icosahedron since it approximates a spherical shape closely) a = area of equilateral triangle (including averaged scalene and isosceles variants).

A set of frequency/area ratios can thus be established to determine which frequencies allow the most effective resonator arrangement while providing the reflector support necessary (variable frequency geodesics will also be explored).

Research is now underway to develop a thin radio-translucent plastic which can transmit audio waves (up to 98% transparent) when in its natural state but becomes reflective (up to 99% acoustically opaque) when a small electric charge is applied. While this latter use may affect the entering radio waves, it is unlikely that astronomical studies will occur simultaneously with acoustic performances (requiring a reflective surface in the geodesic). The plastic must also meet rigid standards of resistance to deterioration from exposure to the sun's ultraviolet light. These triangular membranes should approach a dielectric constant of 4 and a thickness to wavelength ratio of 1/40 creating less than .35 dB loss to signals as they pass through the radome surface. The computer system will control the acoustics of the inner shell (the area between radome and reflector) by various formations of reflective and transparent panels in the geodesic dome thus providing a highly variable acoustic space.

64

Kynar piezoelectric transducers (see #3.B.2 for a more detailed description) will be housed in the main structural elements as microphone/speakers with the hollow beams potentially acting as waveguides and/or resonators (determined during experiments with the test section).

The complementary foundation geodesic (see #2.A.2) follows the same formula as described above. Its principal icosahedron elements will extend out and connect to the radome providing an effective mutual-support system.

B. The Instrument Cabin

1. General Design

The instrument cabin provides two separate locations for equipment. The upper chamber of 2,250 cubic ft. (approximately 10x15x15) will contain the principal components of the transducer I/Os, mixing consoles, amplification units and synthesizer as well as contact with the antenna transceiver located in the feed above the reflector. The lower chamber of 4,000 cubic ft. (approximately 20x20x10) will house the main computer, terminals and disc storage and be acoustically separate from the main cabin to prevent noise interference. This room will also house the central power supply (storage cells, transformers and generators). Researchers will be able to see the reflector and feed through a window set very slightly (14 cm [6"]) above the base of the lowermost plates.

The front entrance and visitor access is provided by a south facing gate in the radome. Stairs up either side of the control cabin provide access to an audience area (no permanent seating) above the electronics control room.

2. Waveguides and Transmission Lines

Because the receiver and transmitter units are located at the feed (sending a 30 MHz center frequency to the main cabin), no extensive waveguides or shielding are necessary. Transmission lines will follow a geodesic "great circle" down to the control cabin area.

C. The Main Receiver

1. Basic Components

The following block diagram describes the "switching radiometer" (Dicke receiver design) for observation of active radio sources:

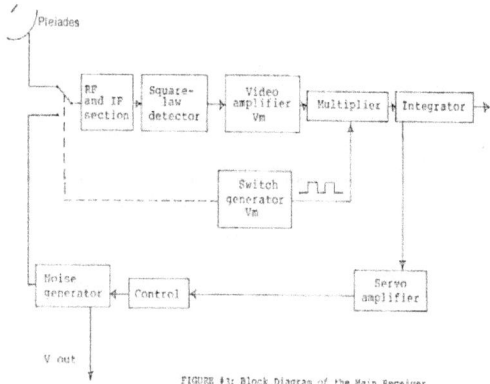

FIGURE #3: Block Diagram of the Main Receiver

Both comparison source and switching frequency can be adjusted in the design. Other more specialized equipment, such as the 8 million channel SETI analyzer, may easily connect to the receiver via the adapters on the main instrument panel.

The radar receiver follows the same basic functions, with the addition of a doppler compensator before the detector stage. Received power will be based on the radar equation:

$$P_r = \frac{P_t\, G^2 \lambda'}{64\pi^3\, R^4} \int_{\pi_2}^{+\pi_2} \lambda^2(\theta)\, \sigma(\theta)\, 2\, \sin\theta\, d\theta$$

where Pr = arriving power, Pt = peak power, G = gain, $\sigma(\theta)$ = radar cross section of the target, and R = energy flux.

2. Read Out Systems

The receiver output will be sampled by analog-to-digital converters, with digital computer processing and control. The computer CRT provides principal data display with soft disc storage available. Access ports enable addition of hard copy printers, strip recorders, magnetic tapes, plotters, and or multichannel analyzers. Both bar graph representation and two dimensional isophote notations will also be available.

D. The Main Transmitter

1. Basic Components

The following block diagram indicates the components of the radar transmitter:

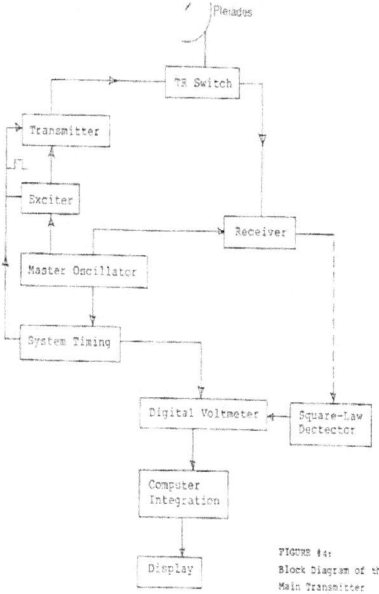

FIGURE #4:
Block Diagram of the
Main Transmitter

The transmitter will normally operate in a pulsed mode at 75 kw average power (500 kw at 3 cm peak power) illuminating the entire reflector surface when directed at the celestial equator. This system will be coupled to the receiver and the read out (display) units.

2. Power Requirements

Operating power requirements will vary depending on the number of incremental elements in use at any given moment. Three power needs must be met: transduction
(energy translation), transmission (radar operation), and environmental (safety
and work lights as well as feed motor control).

The routine receiver and transducer operations will be powered by clean burning
and power efficient generators. Naval (submarine) batteries (or equivalent)
located in the lower cabin compartment will provide electricity for operations
between generator powerups. Radar and performances require separate generators in
order to achieve appropriate peak levels.

Wattage calculations would inevitably vary significantly from exact needs due to
variables such as federal lighting requirements and voltages necessary to power
the largest instrument-plates in the reflector. The final number of generators
necessary for normal and peak power needs will be determined by tests after exact
specifications of electronics and computers are available and during experiments
with the test section (see #2.G).

E. Computer Systems

1. Input

Due to the extraordinary need for simultaneous control of diverse elements
(reflector plates, acoustic panels, radio transmitters and receivers, synthesizer,
Kynar sensors, temperature sensitivity [Kynar pyroelectric sensitivity], lighting,
security, etc.), a "LISP-like" (PROLOG, etc. may be utilized) environment 32 bit
architecture computer will act as the center of control operations. Such capabilities significantly enhance intelligent control of the various functions as well
as offering continuing improvement through ongoing research in artificial and imi-
tative intelligence applications. It will also provide analog to digital conversion and connect to long distance telephone lines near Highway 178 (see Figure
#10) for remote status reports and data transmission. The language FORTH will also
be employed since it is ideal for creating extremely precise code for controlling
many types of electromechanical operations.

2. Data Analysis

A 32 bit "number crunching" computer will be available for use by on-site astronomers, acousticians and composers. When the system is in auto mode this computer
may run SETI programs.

3. Music Control

A third 32 bit machine will handle control of the synthesizers and audio exciter
oscillators. In auto mode this computer will perform transforms of the data
stream. During performances this computer will respond to the Kynar sensors and
modify the audio output according to the user's program.

4. Data Storage

Each computer will have a hard disc for data and a floppy disc drive for programs.
A printer and plotter will be part of the analysis system.

5. Back-up

A duplicate of the cpu of the primary computer will be on hand to prevent interr-
uptions in operation.

A fifth 32 bit machine will be maintained at the *TauCeti Corporation's* office for
software development, and transported to the site for performances. This machine
could also be used for remote operation via modem and the aforementioned telephone
hookup. It should be noted, however, that on-site operation is preferred for reasons of maintenance and security.

F. Clocks and Oscillators

One Hewlett Packard 5065A (Rubidium Vapor) master oscillator and two Hewlett Packard 5100 oscillators (0 to 50 MHz in 0.01 steps) will be made available to astronomers as needed.

G. The Test Section

In order to research the engineering requirements of the structure (reflector surface and support system), a functioning section of the surface (approximately 2,000 sq. ft.) and associated structural elements and resonators will be constructed along with the Gregorian secondary and tertiary reflectors (to be temporarily mounted on a tripod during testing). The plates will follow exactly the design proposed for the full instrument.

Such a test section would provide:

1) exploration of each of the instrument's functions (though with significantly less gathering space);

2) testing of various alloys for the reflector surface, as well as formats for molding and/or machining the spheroid surface;

3) experimentation with various frequency geodesic structures;

4) testing the tuning and shape of the acoustical resonators (see next chapter);

5) demonstration of the aesthetic and scientific usefulness of the instrument (for videotapes, live demonstrations, fundraising, etc.);

6) a research tool for creative work in order that timely and idiomatic experiments take place without the full scale instrument.

Transmitting, receiving and synthesizer equipment could be borrowed to reduce costs and stored in a nearby building rather than in a special instrument cabin.

A site near or on the campus of the University of California at Santa Cruz will be chosen to temporarily support the construction and testing. This test section can then be transported to the Walker Pass site for installation as a member unit of the final structure.

CHAPTER 3
ACOUSTIC CONSIDERATIONS

INTRODUCTION

Recent experiments in acoustics have suggested the immense value of experimentation utilizing large scale instruments. While "paper cone" speakers can crudely simulate sizeable reverberant spaces, particularly in small synthetic circumstances, there remains little question that complexities inherent in *real* dimensional space require complementary facilities. This research station offers a virtually limitless but highly controllable set of sound sources in a massive and equally controllable acoustic theatre with a spherical reflector which acts similarly towards acoustics and electromagnetic energies.

A. Large Scale Instruments

1. The Plates and Electromagnets

A Stradivarius violin, limited in both range and timbre, represents the epitome of musical excellence. The *Pleiades Project* telescope, while offering a wealth of "reproductive" elements (synthesizers, professional speaker systems, Kynar transducers), continues the long tradition of creating "productive" instruments of diversity and complexity for realizing true acoustic and musical sounds. Since almost any non-elastic and potentially vibrating material may be used as surface for the reflector, opportunity for a wide variety of finely tuned and refined instruments exists. Aluminum (and other metals in alloy or pure states), plastics (lucite, PVC, ABS, etc.), and even specially formed glass could serve effectively in uniform surfacing or carefully orchestrated combinations. Vacuum painting (similar to galvanizing) plates not sensitive to magnetic fields with a mixture of epoxy-resin and ferrous metals, will allow induction of vibration from the associated electromagnets.

As described under #2.A.2, the *first* reflector surface, however, will be constructed of specially designed 1/8" to 1/16" diameter galvanized steel (or equivalent alloy) plates supported by 4 to 8 (depending on size of plate) adjustable mounts located at nodal points (to allow controlled oscillations). Throughout developments of these plates (and especially during creation of the test section), great care will be taken to provide extensive timbral index data so that choice of materials, size of plates (vertical, horizontal and by weight) and exact positioning of nodal mounting pins, can be dictated by *musical* criteria. The slight and consistent curve of each plate (to create the sphere of the reflector) should not degrade or significantly influence its ability to function as an efficient and sensitive instrument. One or more electromagnets of appropriate size (varying according to the weight of the associated instrument-plates) will be fixed just below these resonant "idiophones" at distances and locations chosen by on-site measurements with professional Gaussmeters (tests indicate this can be accomplished quickly and easily). A pitch may then be induced into the instrument by exciting the magnet with AC causing proportionately varying attractions.

Tests have proven galvanized steel to be highly diversified in timbre, pitch and resonance and of exceptional musical potential. Unlike traditional speakers, which work in roughly similar ways, metal possesses defined frequency responses with characteristically harmonic and inharmonic series definitions. Thus a single instrument-plate of given size will respond to a fixed set of electromagnetically induced frequencies while almost ignoring others not present in its particular harmonic domain. These steel plates then become extraordinary musical instruments; as bars on a super-sensitive vibraphone with the electromagnets acting as an elegant set of mallets and/or bows capable of exciting one, a few, or all of the large palette of resonant frequencies with a significant array of imposed or endemic envelopes and timbres. A bank of 8+ octave sine wave producing digital oscillators will create an extensive array of timbres through additive synthesis. These may be controlled at the main computer terminals which, after D/A interface and power amplification, will be gated to the appropriate instrument-plates with like frequency partials. Since a given frequency may occur as a fundamental of one instrument-plate while occurring as a fifth harmonic in another (possibly removed by great distance), an immense resource of coupled resonances and acoustic dimensions will be available for research and creative work.

There should be little effect of the electromagnetic induction of the instrument-plates on astronomical observations, since the drivers can be made inoperative during radio-sky research. Concomitantly, phasing of simultaneous plate vibrations with radar broadcasts at similar or harmonic frequencies should provide an extraordinary area of ancillary experimentation.

2. Resonators

The test section will allow experiments with a number of variations of separate resonators set into the reflector substructure through the open triangles of the geodesic. Figure #5 shows a section view of one such arrangement as well as an associated instrument-plate and its support system.

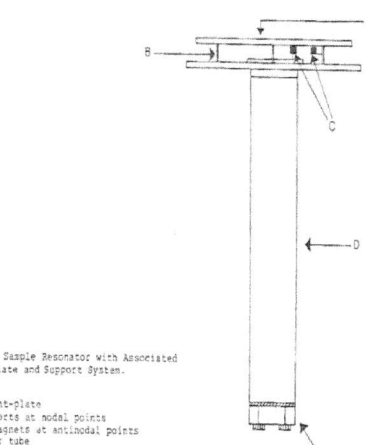

FIGURE #5: A Sample Resonator with Associated Instrument-Plate and Support System.

A = instrument-plate
B = pin supports at nodal points
C = electromagnets at antinodal points
D = resonator tube
E = tuning system

Here, the instrument-plate rests on adjustable pins projecting from steel cross-beams attached to the geodesic. Directly below the plate is an aluminum cylindrical resonator chamber, attached to the same support beams. Exact placement of resonators will be determined by experiments during the construction of the test section.

Adjustable tuning screws allow control of a flat metal disc at the base of each resonator. This disc will sit flush against the curved sides of the cylinder and control the exact volume of the chamber. Tuning of the resonators follows:

$$L = C/4f$$

where L = length of the resonator, C = average velocity of sound at the site (32 degrees Fahrenheit = 1090 ft. per second; 68 degrees = 1,130 ft/sec; 98 degrees = 1,235 ft/sec) and f = frequency of the associated instrument-plate. An "open end correction" of .29 of the inside diameter of the resonator must be subtracted from L to obtain accurate data. Further tuning studies follows:

$$R = 10 \log \frac{I}{I \text{ ref.}}$$

(dB) where R is the value of the response function, I the intensity of the output signal (fixed), and I (ref) a reference signal. When R becomes especially efficient, observed by plotting on a resonance curve, the resonator is considered tuned as in the following:

PLEIADES PROSPECTUS

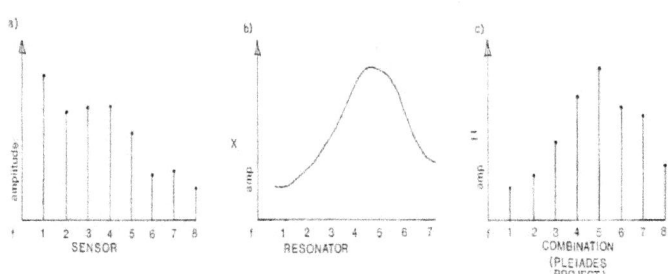

FIGURE #6. Graphs of the Amplitudes of the Harmonic Series of
a) a Sensor Produced Sound; b) Separate Character of
Resonator; c) Sound Produced when Resonator is Attached to
Sensor (a x b).

In this particular analysis, the fifth harmonic is enhanced by the resonator.
Further analysis of the resonators will include studies of nodal (open end) and
antinodal (stopped end) relationships according to the following:

$$f1 = \frac{1}{4L} \; 20.1\sqrt{tA}$$

(L in meters, tA absolute temperature) or half the fundamental frequency of an
open pipe of the same length. Typically the odd partials of a stopped (at one
end) resonator are strongest. The character of the resonator thus matches the
output of the associated instrument-plate.

Figure #5 demonstrates only one possible resonator to be tested on the
instrument-plates. Other types, including Helmholtz spheres, open pipes, and
various "cavity" resonators, will be available for acoustic research and performance. The amplified energy emitted is given by the following:

$$Er = \frac{(1 + \frac{1}{f^2 d^2})}{} \; Eo$$

where Er= resonated energy, f = frequency, d = distance of sound producer from
resonator, and Eo = energy emitted from source without resonance.

Careful attention will be paid during resonator design to balance the acoustic
amplification of the instrument-plates and allow the greatest amount of variation
for control.

The geodesic support structure of the reflector, constructed of hollow steel
cylinders, may also be capable of acting as waveguides for study of acoustic
transmission. The joints could provide opportunity to restrict wave motions to
certain regions or allow these energies to extend throughout large areas of the
superstructure thus producing an almost infinitely variable system for mixing,
matching, phasing and or modulating sounds. Input and output flaired piping could
be added at strategic locations discovered during testing with the instrument-
plates installed. The circumference of the geodesic, for example, would give
excellent output. The test section will provide opportunity to experiment with
this and many other forms of acoustic interaction.

Resonators and waveguides provide extraordinary options for acoustic research and
composition. If both types prove effective, they may also be used in various combinations.

3. Large Scale Acoustics

The sound pressure of spherical waves originating at a point source decreases
inversely with the distance from that source, so that the difference in sound-
pressure level between two points whose distances from the source are D1 and D2 is
given by:

Difference = 20 log D1/D2 dB

The sound level around the Research Station drops 6 db with each doubling of the distance from the source (or 20 db for each tenfold increase of distance).

As a part of the installation, each factor of reverberation (speed, duration and refraction index) will be explored by utilizing the piezoelectric and pyroelectric qualities of the built-in Kynar membranes in the radome (see #3.B.2). These may be used as sensitive microphones and infrared detectors and feed into a system of digital analyzing equipment: various sound level meters including 2 each of Bruel and Kjaer's Digital Frequency Analyzers, Narrow Band Spectrum Analyzers and High Resolution Signal Analyzers.

Experiments with variable intermediate scale acoustic resonances could take place *beneath* the reflector by designing and installing flat horizontal aluminum "walls" to the footings of the lower geodesic. This would create separate rooms and corridors for further tests.

B. Transducers

1. Piezoelectric Wafers

Piezoelectricity or "pressure electricity" takes place when certain crystalline materials change their dimension when subject to an electric field or, conversely, produce electrical signals when mechanically deformed. A variety of piezoelectric acoustic wafers will be available for attaching to the underside of reflector plates as pickups. These provide high resolution timbre and pitch analysis as well as creating a "revised" electronic signal for use in feedback systems or synthesizer operations. These place the main computer in bilateral contact with each reflector instrument-plate insuring accurate tuning, resonance, reflectivity index and associate functions.

2. Kynar Film

Kynar film (trademark: the Pennwalt Corporation) is composed of polyvinylidene fluoride, a polymeric material. This tough, lightweight, flexible and sensitive (frequency roll-off occurs significantly above human thresholds) membrane is both piezoelectric and pyroelectric (capable of producing current based on temperature changes). When thin (6 - 11 microns), such film can act effectively as both microphone *and* speaker and complement the more idiosyncratic virtues of the reflector "idiophones." Kynar is typically coated with nickel (vacuum deposition), though aluminum, tin and various combinations of these metals may also be used. These will be stretched over openings in the structural members of the radome completing the transducer nature of the inner theatre and acoustically paralleling electromagnetic actions in the reflector (though the angle of incidence of the acoustic energy will be more diverse and depend on *which* Kynar membranes are excited). The hollow nature of the reticulated structural elements of the radome may act as acoustic waveguides just as in the lower geodesic (though inaccessibility suggests little potential for adjustment in the upper channels).

Larger timpanic Kynar membranes will surround the rim of the reflector dish (where no distortion of radio signals occurs) and provide a bass frequency range and ability for some tactile interaction with the central system.

All Kynar membranes will be connected to I/O's on the central processor control unit. A wide set of frequencies may be broadcast into the dish and/or picked up (as microphone) from the dish for secondary processing or recording. An extraordinary array of complex interactions between the sensing and playback devices are possible with the large-scale acoustically resonant spherical reflector acting as medium for experimentation and creative endeavor.

3. Audio Electronics

Sounds detected by the pickup transducers will be routed through an automated mixer. Following programmed instructions, these sounds may then be modified, returned to the sensors, recorded, or used to key events from the synthesizers. This facet of the operation will always be under human rather than computer control.

Sounds may be generated electronically by two systems. The *exciter oscillator bank* will consist of banks of oscillators digitally regulated by the music control computer. The outputs of these oscillators are applied to driver transducers placed within the structure as previously described. The *hybrid synthesizer system* will also be controlled by the music computer. It generates a variety of sounds that may be reflected from the structure through the sensors.

Non-electronic sounds (e.g. human voice and other instruments) may be utilized by transduction (by either Kynar or traditional acoustic or contact microphones) to electrical energy (possibly in digital format) for broadcast through the reflecting dish into the acoustic shell of the *Pleiades Project* Research Station.

Sounds emanating from the reflector can be controlled by reflectors attached to the Gregorian feeds. This latter process may only occur during non-telescope-use periods to avoid serious vignetting during radio operations. Standard speakers and microphones will also be placed in this area for recording purposes. Because the installation is so large, video pick-ups and binaural audio monitors may be required during maintenance, tuning and/or performances.

C. Tuning

1. Electromagnet Adjustment

Each instrument-plate will have 2 or more specially designed electromagnets. At least one *drives* the plate by alternately attracting and repelling the galvanized steel with magnitudes equivalent to the output voltages of the digital-to-analog converters. At least one of the other electromagnets *dampens* the plate on cue from the central computer. Foam will be sandwiched between this magnet and the surface of the plate to provide precise control of the amplitude envelope.

The electromagnet drivers and dampers of the instrument may be adjusted to ensure proper focusing at antinodes and exact distances from the instrument-plate surfaces. Careful analysis of nodal factors and frequency (harmonic and inharmonic) will provide data for the computer resulting in accurate placement of the magnets and thus creating a wide set of gradations of attack and decay characteristics. The frequency and timbre index of each instrument-plate may then be tabulated for tuning reference. The emphasis will center on flexibility allowing as many different systems as possible to be incorporated through magnet adjustment and synthesizer control.

2. Harmonicity

Both harmonic and inharmonic relationships prove valuable and will be available in the instrument. Since combined *harmonic* spectra create a macro-resonance allowing the shared frequencies to be enhanced, instrument-plate size and thickness are critical considerations both during construction and prior to performance. The test section allows extensive experimentation with a multitude of options. As well, the research using the "harmonic" spectrum of intervals carried out by Project Director David Cope at the Center for Computer Research in Music Acoustics at Stanford University (using a PDP/10 computer and a specialized digital synthesizer) will be immensely useful.

Inharmonic relationships create "beating" which produces a number of phenomena, not the least of which is a sensation of decay extension and apparent movement within the sound spectra. When the inharmonically related tones originate from separate sources, the beats spin in elliptical patterns through space, from the higher source to the lower one. Each of these and other classifications will be considered in detail during construction of the instrument-plates, placement of the Kynar membranes and module availability in the synthesizer. At all times, flexibility will be a high priority.

CHAPTER 4
THE SITE – FEASIBILITY STUDY

INTRODUCTION

A number of sites for the *Pleiades Project* Research Station were reviewed including the Granite Mountains north of Barstow and the Los Padres National Forest area near Chews Ridge, as well as utilization of pre-existing studies such as the Angel Peak area in the Charleston Mountains north of Las Vegas, Nevada (Drake's 35 meter proposal). While initially promising, each of these sites possessed one or more negative features eliminating it from the list of potential locations. These included lack of accessibility, heavy snowfall, potential seismic activity, no propitious south-facing land shapes, etc. None ever *approached* the extraordinary combination of topographical, geological and environmental advantages of Walker Pass, California, an easily accessed area located approximately 110 miles due north of Los Angeles on the rugged eastern slopes of the Sierra Nevada mountains. The site folds naturally into a half sphere capable of supporting the proposed structures with a minimum of landfill or surface alteration. The "quiet" of electromagnetic frequencies (i.e. relatively free from human intervention) and negligible commercial air traffic (the China Lake Naval Air Station occasionally holds tests in the general vicinity) enhances the potential for radio sky observations. Nearby terrain provides good acoustic absorption rates while also providing a natural "cup" for protection from undesirable local radio noise. The number of cloud-free days, low rainfall and lack of significant snowfall create optimum weather conditions.

A. Topography

Walker Pass is located at 35 degrees 39 minutes 40 seconds latitude, 118 degrees 1 minute 20 seconds longitude on state highway 178 in the southern Sierras, 32 mi. east of Lake Isabella and 8.5 mi. west of state Route 14 (which joins with US 395 7 mi. further north). The closest towns are Jack's Station (2 mi. west of Walker Pass) and Freeman (8.5 mi. east at the junction of state Route 14) each with less than 100 population. The pass lies just north of the National Forest Boundary in the Indian Wells Judicial District. The following maps and photos show the site from a number of viewpoints.

The instrument is designed to fit in a spherical shaped cup of land just south of the peak (elevation: 5,509 ft.) shown below the Pass (reference the Topo map) at an estimated altitude of 5,429 ft. and slightly over 0.4 mi. from the monument next to the road. The location falls just within the Sequoia National Forest. A full survey of the site at 10 ft. topographical interval readings will be compiled and filed and, along with a full geologic report and archaeological survey, will become part of the Environmental Assessment (EA). This will also include a Public Involvement Plan and a planning process with prospective dates for stages of completion.

Drainage from the site flows into an intermittent creek in the canyon directly below the mountain site. This connects with Freeman creek (near Highway 178) which ultimately connects, after crossing the Los Angeles Aqueduct, into the Little Dixie Wash to the southeast of Walker Pass. Like a majority of eastern Sierra runoff creeks, this disappears in a dry lake bed in the general vicinity of the China Lake Naval Air Station. The construction of this project should not affect any natural drainage into this watershed. In fact, careful study suggests that accurate placement of cement footings and the support beams may actually provide "anchors" for the decomposed granite and tend to discourage erosion.

B. Geology

1. Basic Geology

Despite the many cone-like peaks in the neighborhood, there is little evidence of volcanic activity. Geologic maps indicate Pleistocene nonmarine sedimentary depo-

FIGURE #7.1: The negative of a 1972 Air Force satellite photo
of California showing night sky brightness of densely
populated areas (the Pleiades Project is separate and further
protected by mountains):

sits (older alluvium, consisting of slightly consolidated fan deposits, including Quaternary terrace deposits) and an environment of Mesazoic cretaceous granitic rocks (granite, quartz, monzonite and quartz diorite). Close inspection shows that the surface of the slopes is covered with Igneous-Plutonic gravel or decomposed granite. An outcropping of Peridodite indicates a bedrock structure of moderately hard ferromagnisian. Two early 20th century reports indicate magnesite deposits in the general Walker Pass area. The California Desert Conservation Area Plan lists a very low yield Tungsten deposit about 3 mi. to the northeast.

2. Seismic Activity

Walker Pass is 1.5 mi. south of the Pinyon Peak fault which eventually connects with the concealed Sierra Nevada fault zone. This fault slants to the northwest crossing Highway 178 near the Walker Pass Lodge at Jack's Corner (3 mi. from the pass), and again 6 mi. north of Onyx. Since 1934 five earthquakes have occurred near the pass, the most severe of which registered 6.25 on the Richter scale (VIII Mercalli) on March 15, 1946. This produced considerable damage to glass and plaster in the local region (particularly Onyx) and cracked pipes as far away as the City Hall in Bakersfield. Other registered earthquakes include: October 24, 1959

(Richter: 4.2; Mercalli: VI); January 28, 1961 (Richter: 5.3; Mercalli: VI); September 15, 1962 (Richter: 4.9; Mercalli: VI); June 14, 1979 (Richter 4.6; Mercalli: VI).

The Walker Pass area registers as a moderately active seismic area. Earthquakes of similar size to those mentioned above may occur at 8 to 15 year intervals. The lower geodesic support system will provide stability for the reflector during earthquakes. The aforementioned ASTRA computer program will aid in identifying any sensitive structural points in the instrument. Appropriate antishock buffers will be designed if necessary to insure safety and stability during earthquakes.

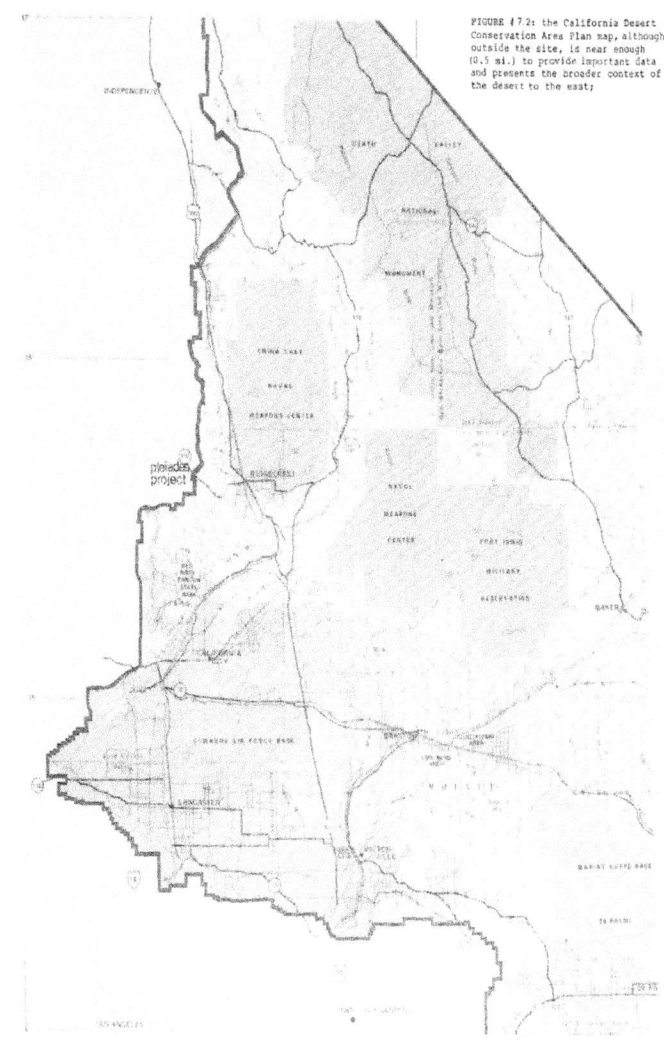

FIGURE # 7.2: the California Desert Conservation Area Plan map, although outside the site, is near enough (0.5 mi.) to provide important data and presents the broader context of the desert to the east;

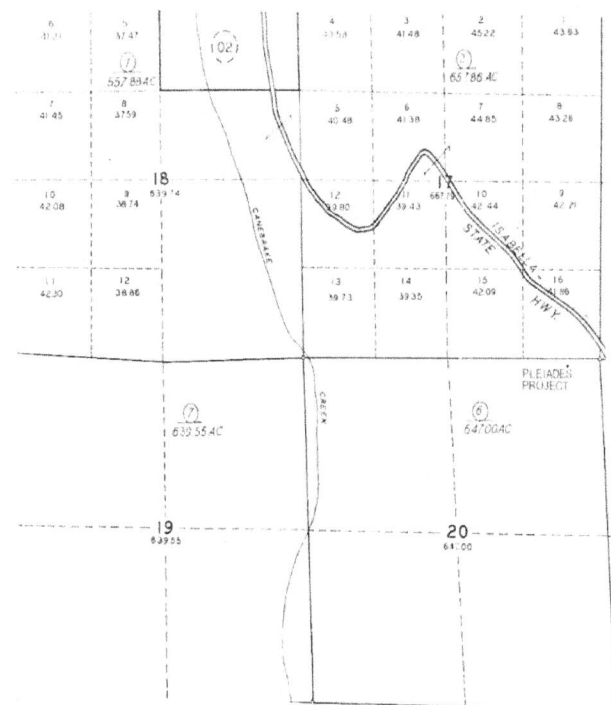

FIGURE 17.3: the Kern County Assessor's Map shows location of nearby private and BLM land:

C. Climate, Altitude and Wind Considerations

Average temperatures for Walker Pass range from 26 degrees Fahrenheit in mid-January to 86 degrees in mid-July. Extremes, while not officially registered, are estimated at near 0 degrees in winter and 98 degrees in summer. Careful testing to minimalize surface deviations due to temperature expansion and/or contraction will take place prior to final testing and installation (see #2.A.2 and the test section in #2.G).

Walker Pass receives an annual sun/cloud ratio of 4:1 (288 days average sun to 72 overcast or stormy weather). Since high frequency observations require cloudfree conditions, this factor is significant to site selection. The following map shows that Walker Pass falls within the clear day (0 - 3 tenths Mean Sky Cloud Cover) band of the western U.S. See Figure #8.

Rainfall varies from 8 to 19 inches per year. Snowfall (due to the more than mi. high altitude) occasionally closes the Pass in mid-winter usually due to icy road conditions rather than significant snow pack.

Current analysis of wind conditions in the Walker Pass area (from the *Handbook of Geophysics and Space Environments*, N. Y.: McGraw Hill, 1966) indicates 5 minute peak winds at less than 60 mph (isopleths of 1% of 25 year maximum 5 minute winds in mph at 50 altitude). On-site measurements in both fair and poor weather conditions indicate westerly wind directions at 30 mph maximum. The proposed site on the south face of the mountain receives minimal crosswinds due to the southeast direction of the air flow tunnel. Summer storms originating in the Gulf of Cali-

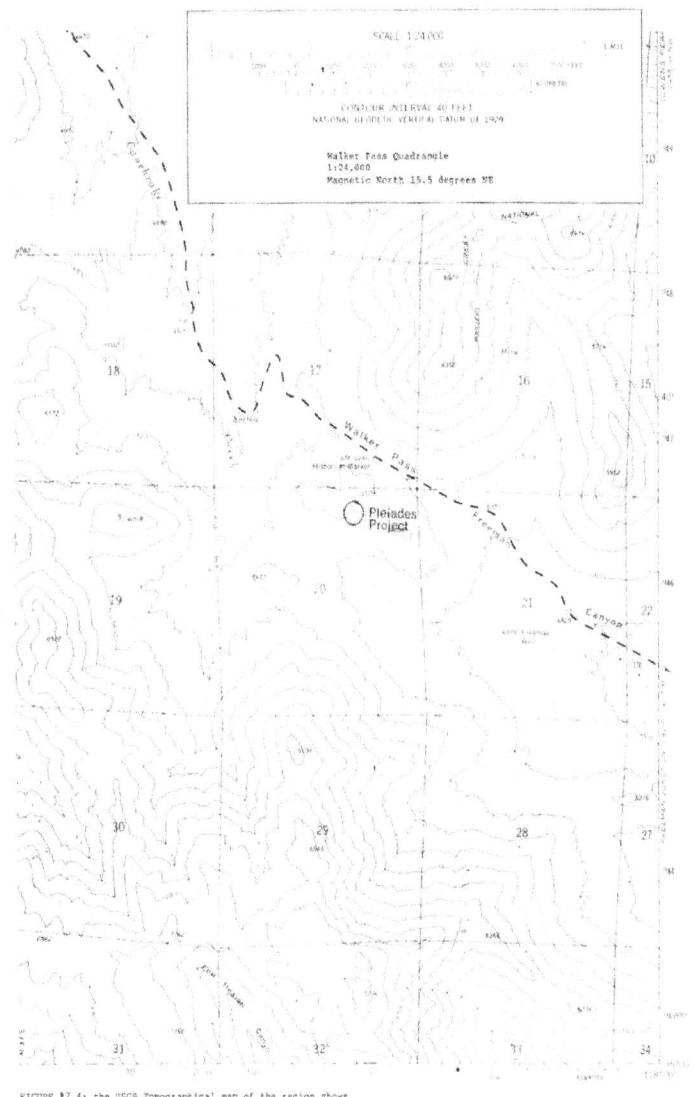

FIGURE 17.4: the USGS Topographical map of the region shows
the benchmark altitude of the Pass at 5,245 ft. above sea
level;

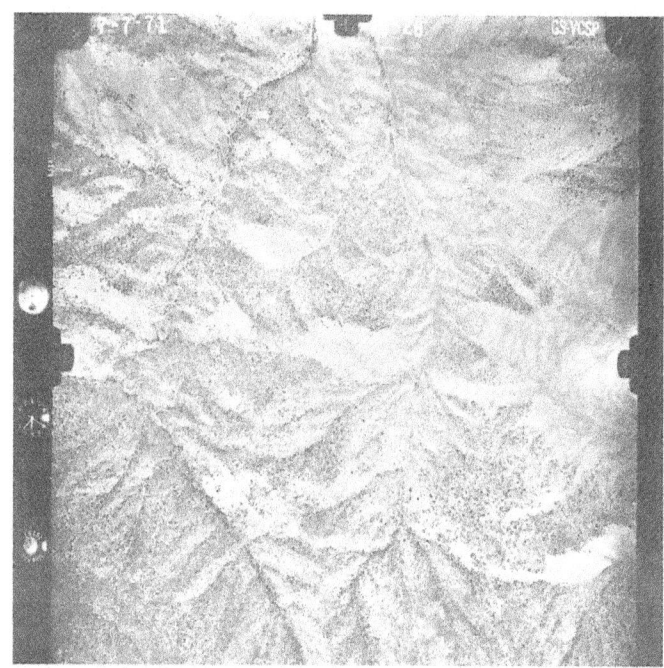

FIGURE #7.5: a USGS aerial survey photo of Walker Pass;

fornia (notably during August and early September) may occasionally surpass predicted wind velocities and provide electrical displays. The radome, however, will provide more than adequate protection for the instrument from these events.

D. Environmental and Cultural Impact

1. Vegetation

The pass is surrounded by stands of Joshua trees native to the area of the western Mojave Desert. These *Yucca brevifolia* often grow to heights of 25 ft. or more, live in excess of 200 years and possess trunks as hard as Douglas firs.

Utah Junipers (Juniperous utahensis) occupy the peak area and the northern slope. Sagebrush (Artemesia tridentata), in a variety of forms (e.g. bladder sage: Salazaria mexicana), covers much of the region. The Huckleberry Oak (Quercus vaccinifolia) and California Buckwheat (Eriognum fasciculatum) also occur but in lesser numbers. Chamise (Adenostoma fasciculatum), Bitterbrush (Purshia tridentata), Rabbit Brush (Chrysothamnus nauseiosus) and Golden Yarrow (Eriophyllum confertiflorum) occur in some areas on the northern face and on the peak itself.

The southern slope of the site is relatively barren except for approximately 5 small Junipers, a stand of Yucca (significantly below the site), and a plethora of sage and buckwheat.

FIGURE #7.6: a view from north of the pass (the proposed site is behind the central mountain - note the damage from a 1972 fire to the left of the peak);

FIGURE #7.7: a photo from the peak above the site towards the south;

FIGURE #78: a photo from below the site (looking north) showing the natural cup of land for the instrument.

2. Wildlife

The raised mound burrows of the broad-footed mole (Scapanus latimanus) can be seen in spots near the summit of the site and the lair of a tentatively identified kit fox (Vulpes macrotis) was found approximately halfway up the northern face. Evidence of both the Coast Horned Lizard (Phrynosoma coronatum) and the Striped Racer (Masticophis lateralis) appeared in one site though neither was verifiable. Rabbit droppings have also been found at various locations near the summit. The United States Forest Service and Bureau of Land Management grazing permits have allowed both sheep and cattle access to the area.

FIGURE #8

Mean Annual Number of
Clear Days at 19 Southwestern
U.S. Weather Bureau Stations.

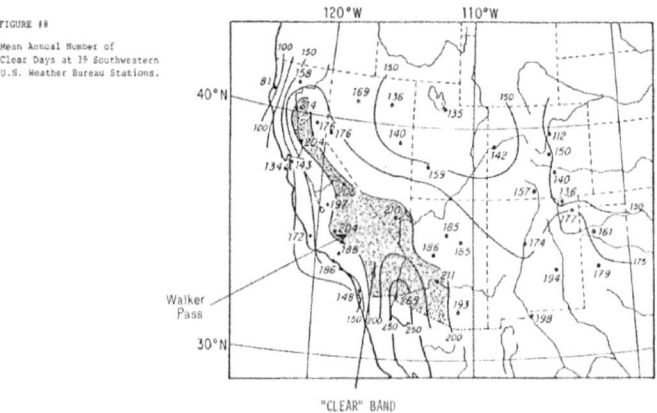

The California Desert Conservation Area Plan, an extensive EIR of the area beyond the eastern border of the Sequoia National Forest (see Topo map for clarification), lists the "yellow-eared pocket mouse" as a potentially endangered species (extending to Highway 395). However, the area so defined begins at elevations significantly below the site and much to the east. The Plan also indicates the Golden Eagle (Aguila chrysaetos) as a sensitive species and dedicates over 2.5 million acres of protection from hunters. These large birds (often over 7 ft. wingspread in adulthood) tend to nest in the higher mountain crags or near grasslands atypical of those of the site and the CDCAP indicates occupancy at elevations *below* the currently proposed Walker Pass area. No other plants or wildlife in the immediate region are currently designated as endangered species. Hence, construction should not pose any problems for threatened species.

3. Archaeological Resources

The amerinds of the Walker Pass region belonged to the Shoshonean linguistic group and were most directly related to the Chemehuevi of southeastern California (though culturally different in many ways). Called *Kawaiisu*, their tribal differentiation places them somewhere between the Shoshone and Yokuts (principally Yauelmani) with their principal occupation in Walker basin and the Caliente Creek watersheds. Walker Pass was not a critical interior location; however, both the Chemehuevi and Tubatulabal apparently used the pass for eastern Sierra crossings as did the Coso to the northeast (though this latter tribe had many other choices).

The *Kawaiisu* left occasional pictographs in the Isabella and Kern regions to the immediate west of Walker Pass though none occur on or near the proposed installation. Recent communication with Robert Schiffman of Bakersfield College, recognized by the USFS and Sequoia National Forest as the archaeological authority on this region, indicates several small lithic scatters (some with milling features) near the campground but no recorded pictographs or other sites within a 2 mi. radius of Walker Pass. Additionally, Dr. David Whitley, Chief Archaeologist at the Institute of Archaeology at UCLA and an expert on this region, has notified the project that he knows of no sites near Walker Pass and is not aware of the area holding any special significance for the Kawaiisu, Chemehuevi, Coso or Tubatulabal.

G. Land Ownership and Access

1. Ownership

Walker Pass is located near the northeast boundary of the 1.8 million acres of the Scodie Mountain portion of the Sequoia National Forest. This area, originally under a "Roadless Area Review Evaluation" (see final Environmental Statement: RARE II, USDA Forest Service Document FS-325 and California Supplement), has recently (September, 1984) been designated a "further planning" region to study its potential as a designated "wilderness." Since adoption of such a classification requires maintenance of the natural character of the site, the *Tauceti Corporation* has formally requested "non-wilderness" status for the Scodie Mountains or, if "wilderness" status is given, that the acreage directly south of Walker Pass be exempted. Robert D. Addison, District Ranger in Kernville, has invited input from the *Pleiades Project* to the planning process. The deciding document, the Forest Land Management Plan, should be available in draft form early in 1985.

The Sequoia National Forest is administered by the United States Forest Service, a branch of the U.S. Department of Agriculture whose Handbook provides the rules and restrictions pertaining to the use of National Forest Property:

> "The Secretary of Agriculture is authorized, under such regulations as he may make and upon such terms and conditions as he may deem proper, (a) to permit the use and occupancy of suitable areas of land within the national forests, not exceeding eighty acres and for periods not exceeding thirty years, for the purpose of construction or maintaining hotels, resorts, and any other structures or facilities necessary or desirable for recreation, public convenience, or safety; (b) to permit the use and occupancy of suitable areas of land within the national forests, not exceeding five acres and for periods not exceeding thirty years, for the purpose of constructing or maintaining summer homes and stores; (c) to permit the use and occupancy of suitable areas of land within the national forest, not

exceeding eighty acres and for periods not exceeding thirty years, for the purpose of constructing or maintaining buildings, structures, and facilities for industrial or commercial purposes."

   - Agriculture Handbook No. 453-1974 (Act of March 4, 1915 [38 Stat. 1101, as amended; 16 U.S.C. 497])

A preliminary review process is underway between the *Tauceti Corporation* and the Cannell Meadow Ranger District in Kernville. A Resource Officer of the Sequoia National Forest has visited the site with Director David Cope and found no immediate or obvious problems with site utilization.

The Research Station has been tentatively designated by the USFS as an "electronics site" requiring lease at 0.2% of the estimated value of the electronics equipment in the structure (approximately $800 per year).

Precedence in the use of National Forest property for observatories is provided by the Monterey Institute for Research in Astronomy (MIRA) in the Los Padres National Forest and for radio antennae, the Continental Microwave site in the Sequoia National Forest 6.2 mi. southwest of Walker Pass on a 6,599 ft. peak.

2. Road Access

Highway 178 is the principal paved route through Walker Pass. A gravel road connects to this highway approximately 1.2 mi. to the southeast from Walker Pass and turns immediately due west toward the Pass. Travel of about 1 mi. on an occasionally rough (primarily due to erosion from "flash" storms) and unimproved road places one almost directly south of the proposed site. Approximately two-thirds of the road from Highway 178 falls under the jurisdiction of the United States Bureau of Land Management (BLM; Department of the Interior) and is currently designated a part of the California Desert Conservation Area Plan (CDCAP). This plan defines the area the road crosses as "Class L" which allows for construction of communication sites, removal of vegetation and new road construction, while restricting agriculture and waste disposal. It also requires strict management of all water resources according to the Clean Water Act: Section 208, EO 12088. Preliminary communication with the BLM District Office in Bakersfield has begun. Request for a review process will be made in the near future. The last third of the road falls under the jurisdiction of Sequoia National Forest which regards the land on which it resides as Zone D: Off Road Vehicle traffic is acceptable provided it causes no resource damage or user conflict. Accurate distances and access to the site appear in maps under #4.A.

3. Trails

The Pacific Crest trail (PCT) is tangential to the Walker Pass campground at a point approximately 0.16 mi. to the east of the first turn up the western approach to the pass and about 0.6 mi. west of the pass itself. From there it moves north and then parallel to the State Highway, crossing Highway 178 at Walker Pass. The instrument will not be visible on this portion of the trail. Only hikers south of the pass, at distances of 2.5 miles, could see the Research Station; a view which would include the China Lake Naval Air Station and much of the valley and dry lakebed beyond. It is highly unlikely that sound will reach any portion of the PCT, even during the loudest performances, due to intervening mountains and/or extreme distances (see #4.H.1).

A 0.3 mi. fire break (of roughly 40 degrees grade) begins at the junction of the trail and the campground and culminates in dense wood approximately 0.2 mi. from the site.

F. Impact of Site Occupation

1. Audience Requirements

Due to the remote location, it is not anticipated that large audiences will attend performances. The *Tauceti Corporation* (in cooperation with the Sequoia National Forest) will carefully monitor audience potential.

2. Water and Sanitation

The initial proposal calls for construction of an on-site well to provide water for researchers and visitors to the station. Should this not be possible, wells occur at two nearby locations: a) Freeman Well (with significant water flow) is 0.3 mi. from Highway 178 along the gravel road about 0.7 mi. east of the site at approximately 4878 ft. altitude and consists of a 30 ft. high windmill and a 20 ft. diameter circular (6.5 ft. high) reservoir; b) the campground (0.76 mi. from the site to the west) has an active spring (approximately 0.12 mi. due west of the parking circle) and a spigot providing suitable drinking water. Water could be hauled from these sites to a moderate size tank near the living quarters in the Research Station.

The initial proposal also calls for a septic tank/leach system to be installed for sewage disposal. If this proves unsuccessful, stainless steel containers could be hauled from the site to the campground (0.76 mi. to the west) which has two high grade pit toilets which are regularly pumped and reset with chemicals.

3. Housing

A medium size trailer located below the reflector yet inside the radome will provide housing for researchers and assistants. Temporary housing for non-researchers could be maintained at the recently (1980) completed Walker Pass Campground (0.76 mi. west-northwest of the site or 3.5 mi. by road). This campground is maintained by the BLM and accepts moderate and even large sized trailers, RV's and motor homes. There are no "stay limits" on these vehicles since the site is often used as a point of departure for hikers on the Pacific Crest Trail (a PCT trailhead). As an alternative, a small area along the BLM portion of the proposed approach road (see Figure #10) could be constructed and maintained for this purpose.

G. Aesthetic Considerations – Visual

The USFS has designated the Walker Pass area as:

$$\frac{MG1B}{PR}$$

which translates as MG = middleground; 1 = high sensitivity; B = common class; PR = partial retention defined as "...activities remain visually subordinate to the characteristic landscape..." (from USDA *Forest Service* "National Landscape Management" Vol. #2 - Agriculture Handbook No. 462).

The instrument's design avoids conspicuous public attention by its south face construction away from the nearest highway. A single access route makes the instrument available to those interested in either its visual or aural impact. Each approach route (by foot) has been plotted to achieve an ambience with Walker Pass. Great care has been taken to integrate the Research Station with the pyramid-like peaks in the area behind and to either side of the site. The passive acoustic resonances should project little beyond 100 ft. thus sounding only for those who make the effort to approach the instrument.

The Research Station will not create significant horizon alteration from any point aside from that directly below its dish. The dome should be inconspicuous and generally hidden by the trees in the area.

H. Aesthetic Considerations - Aural

1. Performances

During the loudest performances and with the upper geodesic panels in a translucent mode, the large scale instruments in the Research Station could produce sound levels at 20" distance approximating 99 dB maximum peaks on rare occasions. These sounds should drop 20 dB for each tenfold increase in distance (barring obstacle interruption). In optimum situations, the following should suffice as average sound pressure levels during performance of the loudest events.

distance from site	dB level
20"	99 dB
200" (16')	79 dB
2000" (166')	59 dB (normal speech level)
20000" (1666')	39 dB
200000" (16,666')	19 dB (whisper)

Since the originating source does not exceed the standard off-highway motorcycle noise limits (see Off-Highway Motorcycle and All Terrain Stationary Test Manual of the Motorcycle Industry Council), no undue acoustic pollution should be generated on or around the site. Figure #9 provides a "topacoustical map" of the area based on the above information:

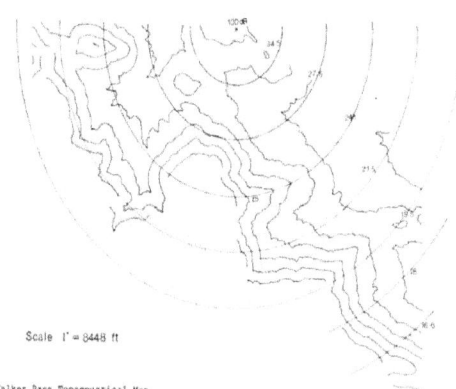

Scale 1" = 8448 ft

FIGURE 89 : A Walker Pass Topacoustical Map.

Wind and temperature variations in the atmosphere around the instrument will modify the distribution of energy by bending the waves from their rectilinear paths. The speed of sound in still air is given by:

$$V = \sqrt{1.40 \ P/d}$$

where V = velocity, P = atmospheric pressure, and d = density. The speed is therefore inversely proportional to the square root of the density of the air and therefore directly proportional to the square root of its absolute temperature. Thus, if the air is in motion, or if the temperature changes, the sound speed is altered. Frequency and wavelength can then be calculated according to the following formulae:

$$f = \frac{20.1}{\lambda} \sqrt{Ta} \qquad \lambda = \frac{20.1}{f} \sqrt{Ta}$$

Temperature differences cause refraction of sound waves. Commonly air temperature decreases with the altitude above the earth's surface. In this case, the upper portions of sound waves originating at the sound source are retarded in relation to the lower portions and consequently the wave front bends upward. Conversely, if the temperature increases with altitude (inverted temperature gradient), then the upper waves travel faster and the wave front bends downward. These effects may cause rare exceptions to the 20dB rule, but the likelihood of any significant variation is nonexistent.

2. Transducer Performance

During acoustic interpretations of stellar phenomena, sounds will be confined to extremely low levels (averaging 50 dB at a 1 ft. distance from the source) and should be inaudible beyond a 100 ft. perimeter outside the radome.

CHAPTER 5
CONSTRUCTION, MAINTENANCE AND MANAGEMENT

INTRODUCTION

This Research Station will be constructed, operated and maintained by the *Tauceti Corporation*, a nonprofit scientific, artistic, and educational organization. The Officers and Board of Directors will make the instrument available through routinely published invitations for applications for composition and research.

A. Construction Operation

1. The Tauceti Corporation

The *Tauceti Corporation* is a nonprofit, tax exempt organization located in the State of California and dedicated to support collaborative and professional interdisciplinary research in the arts and sciences. Its Federal Internal Revenue Service Employer Identification Number is: 77-0030862. Its Articles, By-Laws and state and federal Tax Exempt Forms are available for public observation at the main office: 317 Nobel Drive, Santa Cruz, California, 95060.

2. Organization

The Officers of The *Tauceti Corporation* consist of a President, Secretary and Treasurer who report to the Board of Directors. A group of advisors serve to coalesce the designs of the various acoustic and electromagnetic functions of the electromagnetic-acoustic instrument. These include architectural designers, radio astronomers, electro-computer experts and a technical advisor. Business advisors (including legal counsel, a fiscal consultant, and a public relations manager) will serve to coordinate the EIR, fund raising and publicity aspects of the project. Other advisors (e.g. engineers, designers) will be sought for particular situations.

3. Construction

The *Tauceti Corporation*, its Board of Directors, Officers and Advisors, will oversee all construction operations. Contracted work (especially road improvements and hauling) may be augmented with support from qualified volunteers.

It is anticipated that all antenna structures and cabin equipment will be transported overland (helicopter installation remains an option).

None of the grades exceed standard ceilings with most significantly below accepted limits. Great care has been taken to avoid tree removal and to prevent inordinate damage to area sagebrush and buckwheat. Ground cover not on the road itself should return to its present general state within 5 years. The last (dotted line) switchbacks will only be necessary if the hoisting of reflector and radome materials is unsuccessful.

B. Protection/Security

1. Theft/Destruction/Vandalism

The radome covering the instrument should provide generally secure quarters for researchers and equipment. Entrances will be locked except during announced performances. The low visibility of the structure should prevent extensive curiosity from passing motorists. As well, the USFS indicates extraordinary success with their campground monitors and suggest that site *occupation* is the most effective deterrent to theft, destruction and/or vandalism.

The following map presents a proposed road for use during construction:

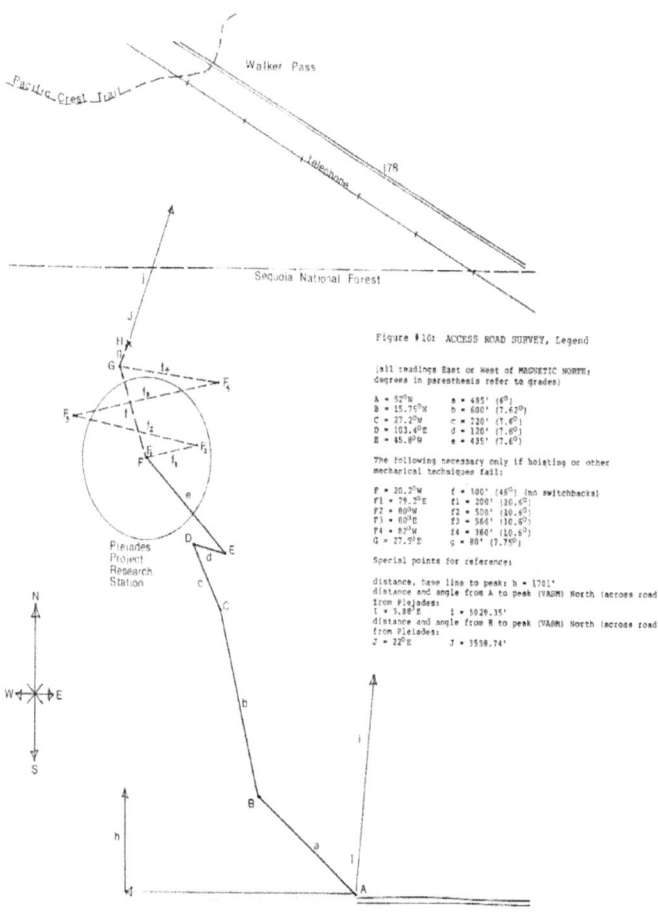

Figure #10: ACCESS ROAD SURVEY, Legend

(all readings East or West of MAGNETIC NORTH; degrees in parenthesis refer to grades)

A = 52°N		a = 485' (6°)	
b = 15.75°N		b = 600' (7.62°)	
C = 27.2°W		c = 220' (7.6°)	
D = 103.4°E		d = 120' (7.6°)	
E = 45.8°W		e = 435' (7.6°)	

The following necessary only if hoisting or other mechanical techniques fail:

F = 20.2°W	f = 100' (46°) (no switchbacks)
F1 = 79.5°E	f1 = 200' (16.6°)
F2 = 80°W	f2 = 550' (10.6°)
F3 = 80°E	f3 = 560' (10.6°)
F4 = 83°W	f4 = 360' (10.6°)
G = 27.5°E	g = 80' (7.75°)

Special points for reference:

distance, base line to peak: h = 1701'
distance and angle from A to peak (VABM) North (across road) from Pleiades:
i = 5.88°E   i = 5029.35'
distance and angle from B to peak (VABM) North (across road) from Pleiades:
J = 22°E   J = 3550.74'

When the Research Station is not occupied for reasons of scientific and/or artistic research, a local (Ridgecrest) agency will be hired on a contractual basis (by the *Tauceti Corporation*) to visit the site regularly and check for developing natural problems or vandalism. Clear postings in and around the instrument area should impede acts of vandalism though visitor movement around the site is expected.

2. Personal Injury

Appropriate liability insurance will be secured before construction by the *Tauceti Corporation*. Visitors to the site during and after construction will be protected by agreement of the *Tauceti Corporation* and the Sequoia National Forest (statutes of the U.S. Department of Agriculture).

C. Maintenance

The *Tauceti Corporation* will be responsible for full maintenance of the research station instrument through an endowment the express purpose of which is to insure ongoing research and performance. Sub-contracts with regional electronics firms will guarantee inspection on a regular basis and repair with nominal delays.

D. Management

1. The Endowment

As stated in Chapter #1, a major goal of the *Tauceti Corporation* will be the acquisition of sufficient funds to establish an endowment to support the ongoing research and composition.

2. Application Procedure

Any individual may, following established procedures, apply to the *Tauceti Corporation* for use of the instrument. Priority will be given to those with credentials and experience in areas relevant to the instrument's use. A *Users Manual* will be made available to all researchers providing information regarding the function and operation of the equipment.

3. Selection Process

A selection committee formed by the *Tauceti Corporation* will solicit and review applications and (within set guidelines and available funds) grant use periods. All accepted applicants will be required to follow specific rules and regulations as set forth in contracts of the corporation.

E. The Schedule

1. The Phased "Design-to-Construction" Schedule

Scheduling an actual completion date is impossible since two significant variables, site acquisition and the securing of funding, remain unpredictable. It is probable, however, that once these two needs are met the station can be built in 10 months or less. The following represents a general overview of the proposed schedule:

End of Month	Activity
1	full staff assembled
	final designs accepted
2	separate engineering prospectus
	completed
3	test section begun
4	section testing
5	prefabrication of plates begins
6	site preparation (and approach)
	begins
7	transport of plates
8	assembly and construction on-site
9	installation of electronics and
	computers
	security tests
10	final tests
	initial operations begin

2. The Research and Performance Schedule

Research and performances will take place after approval by the *Taucet/ Corpora-tion* at least 6 months in advance of proposed dates. The following provides examples of applications:

A. PROJECT TITLE: A study of organic molecules in the constellation of Pleiades

ABSTRACT OF REQUEST:

The nebula surrounding the youthful stars of the Pleiades star cluster (23 - 24 degrees declination) will be studied for signs of organic molecules with special concern for methanol (CH_3OH at 24.93 GHz) and Formaldehyde (H_2CO at 28.97 GHz). Analysis of these molecules and others, possibly yet to be discovered in space, could prove of extraordinary importance to astronomical research. Little work has been done in the range of 20-60 GHz to date since no extant single instruments have such capability.

DISSEMINATION: Publication of general results in various refereed journals.

B. PROJECT TITLE: "Earth Diver:" a musical composition.

ABSTRACT OF REQUEST:

November 17 (the culmination [midnight crossing of the zenith] of the star cluster Pleiades) offers extraordinary opportunity to celebrate the Pleiades' legend found in local Kawaiisu *Coyote* myths. We plan to use the resonance of the station as well as interplanetary space for a new work by employing the "radar" nature of both sound and electromagnetic energy. Sections of the composition will be reflected from the full moon (in Taurus on the proposed date) with 2.57 second delay, Mars (in Capricorn) with 8.3 minute delay, and Jupiter (in Aquarius) with a 1 hour and 6 minute delay, simultaneously with reverberation below the variably acoustic radome. Spatial modulations (movement of sound sequentially across and around the surface of the reflector) and other techniques may be employed as a result of incremental analysis.

DISSEMINATION: These explorations will be integrated into a collaborated creative work and then videotaped for later TV and theatre release.

C. PROJECT TITLE: "Phase Relationships: Electromagnetic and Acoustic Energies"

ABSTRACT OF REQUEST:

Initially an unmodulated 10 KHz radio signal will be broadcast into the reflector via the radar transmitter. The electromagnets below the instrument-plates will simultaneously emit complementary (harmonically resonant) and modulated frequencies while slowly shifting out of phase. Observations (mainly in the Kynar membranes since they are sensitive to radio as well as audio waves) will take place of all manner of harmonic and inharmonic sideband activities and associated series. In reverse, an unmodulated audio (Kynar) 10 KHz signal will be broadcast from the radome concomitantly with a similar but modulated frequency radar emission from the feed for parallel observations. White noise may also be used in separate experiments.

DISSEMINATION: Publication as an IEEE research monograph.

3. The Telescope—Transmitter Schedule

The RF transmitter will operate per licenses from (and according to regulations of) the Federal Communications Commission. Radar broadcasts (for exploration of comets, the moon and other local objects) will be interfaced with NASA and the nearby Mars (JPL at Goldstone; north of Barstow) and Owens Valley antennae as well as the nearby China Lake Naval Air Station. When utilized for creative endeavors, all operations will be coordinated with appropriate national and international scientific organizations.

BIBLIOGRAPHY

(a select list of referenced sources)

Askill, John. Physics of Music Sounds. New York: Van Nostrand, 1979

Auld, B. A. "Wave Propagation and Resonance in Piezoelectric Materials." Journal of the Acoustical Society of America: December, 1981

Benade, Arthur. Fundamentals of Music Acoustics. New York: Oxford University Press, 1976

Berkowitz, Ami and Eckart Kneller. Magnetism and Metallurgy. New York: Academic Press, 1969

Brown, R. Hanbury and Sir Bernard Lovell. The Exploration of Space by Radio. New York: John Wiley and Son, 1962

California Desert Plan. Washington: Dept. of the Interior, 1980

Chamberlain, Hal. Musical Applications of Microprocessors. Rochelle, New Jersey: Hayden Books, 1981

Chowning, John. "The Simulation of Moving Sound Sources." Journal of the Audio Engineering Society: 2-6 (1971)

Cope, David. "A View of Electronic Music." db The Sound Engineering Magazine: Vol. 9, No. 8: August, 1975

------------. New Directions in Music (4th ed.). Dubuque, Iowa: Wm. C. Brown Co. Publishers, 1983

Culver, Charles A. Musical Acoustics. New York: McGraw-Hill, 1956

Drake, Frank. NAIC Proposal to Conduct Cost Studies of a 35-meter Millimeter Wavelength Telescope. Ithaca: New York: NAIC, 1979

------------. Users Manual for the Arecibo Observatory. Ithaca, NY: NAIC, 1979

Evans, John V. and Tor Hagfors (editors). Radar Astronomy. New York: McGraw-Hill, 1968

Everest, F. Alton. The Master Handbook of Acoustics. Blue Ridge Summit, PA: Tab Books, 1981

Findlay, J. W. and S. von Hoerner. A 65-meter Telescope for Millimeter Wavelengths. Charleston, VA: NRAO, 1972

Fuller, R. Buckminster. Synergetics. New York: Macmillan, 1982

Garfinkel, A. P., R. A. Schiffman and K. R. McGuire. Archaeological Investigations in the Southern Sierra Nevada: The Lamont Meadows and Morris Peak Segments of the Pacific Crest Trail. Bakersfield, Ca: Bureau of Land Management, Cultural Resources Publications in Archaeology, 1982

Giddis, Albert R. Reflector Antennas for Radio and Radar Astronomy. Palo Alto: Philco Corporation, 1961

Grey, John. "Multidimensional Perceptual Scaling of Musical Timbres." Journal of the Acoustical Society of America: May, 1977

Groves, Ivor D., Editor. Acoustic Transducers. Stroudsburg, Pa: Hutchinson Ross Publishing Company, 1981 (Benchmark Papers in Acoustics Vol. 14)

Hall, Donald. Musical Acoustics. Belmont, CA: Wadsworth Publishing Co., 1980

Hew, J. S. The Radio Universe (3rd edition) Oxford, England: Pergamon Press, 1983

Hulten, K. G. Pontus. The Machine. New York: The Museum of Modern Art, 1968

Hutchins, Carleen M. Musical Acoustics (Vol. 1 and 2). Stroudsburg, Pa: Hutchinson Ross Publishing Company, 1981 (Benchmark Papers in Acoustics Vol. 5 "Violin Family Components," and Vol. 6 "Violin Family Functions")

Kent, Earle L. Musical Acoustics. Stroudsburg, Pa: Hutchinson Ross Publishing Company, 1982 (Benchmark Papers in Acoustics, Vol. 9)

Kawakami, F. and K. Yamaguchi. "Space-ensemble Average of Reverberation Decay Curves." Journal of the Acoustical Society of America: October, 1981

Kluver, Billy, Julie Martin and Barbara Rose, editors. Pavilion, Experiments in Art and Technology. New York: E. P. Dutton and Co., 1972

Knudson, Vern and Cyril Harris. Acoustical Designing in Architecture. New York: John Wiley and Sons, 1950

Kock, Winston. Sound Waves and Light Waves: The Fundamentals of Wave Motion. New York: Doubleday and Co., 1965

Kraus, John. Radio Astronomy. New York: McGraw-Hill, 1966

Kroeber, A. L. Handbook of the Indians of California. New York: Dover Publications, 1976 (original in 1925)

Kynar Piezo Film Technical Manual. King of Prussia, Pennsylvania: Pennwalt Corporation, 1984

Lee, Chun P. and Taylor G. Wang. "The Acoustic Radiation Force on a Heated (or Cooled) Rigid Sphere - Theory." Journal of the Acoustical Society of America: January, 1984

Malecki, I. Physical Foundations of Technical Acoustics. London: Pergamon Press, 1969

Mar, James and Harold Liebowitz (editors). Structures Technology for Large Radio and Radar Telescope Systems. Cambridge, MA: MIT Press, 1969

Marcus, Michael A. "Ferroelectric Polymers and their Applications" Pennsylvania State University: 5th Annual International Meeting on Ferroelectricity, 1981

Mathews, Max. The Technology of Computer Music. Cambridge, MA: MIT Press, 1969

Mehl, James B. "Acoustic Resonance Frequencies of Deformed Spherical Resonators." Journal of the Acoustical Society of America: May, 1982

Miller, Harry B., editor. Acoustical Measurements. Stroudsburg, Pa.: Hutchinson Ross Publishing Company, 1982 (Benchmark Papers in Acoustics, Volume #16)

Northwood, Thomas D. Architectural Acoustics. Stroudsburg, Pa: Hutchinson Ross Publishing Company 1982 (Benchmark Papers in Acoustics, Vol. 10)

Oliver, Bernard and John Billingham. Project Cyclops. Ames Research Center, CA: NASA CR 114445, 1971

Pain, H. J. The Physics of Vibrations and Waves. New York: John Wiley and Sons Ltd., 1983 (third edition)

RARE II. Washington: USDA - Forest Service (FS - 325), 1979

Reed, C. A. L. and K. F. Sander. Transmission and Propagation of Electromagnetic Waves. Cambridge: Cambridge University Press, 1978

Roads, C. "Artificial Intelligence and Music." Computer Music Journal: Vol. 4, No. 2, Summer 1980

Roederer, Juan. Introduction to the Physics and Psychophysics of Music. London: The English University Press, Ltd., 1974

Rossing, Thomas D. The Science of Sound. Reading, Mass.: Addison-Wesley Publishing Company, 1982

Stanley, R. C. Light and Sound for Engineers. London: Thomas Nelson and Sons, 1968

1.

2.

3.

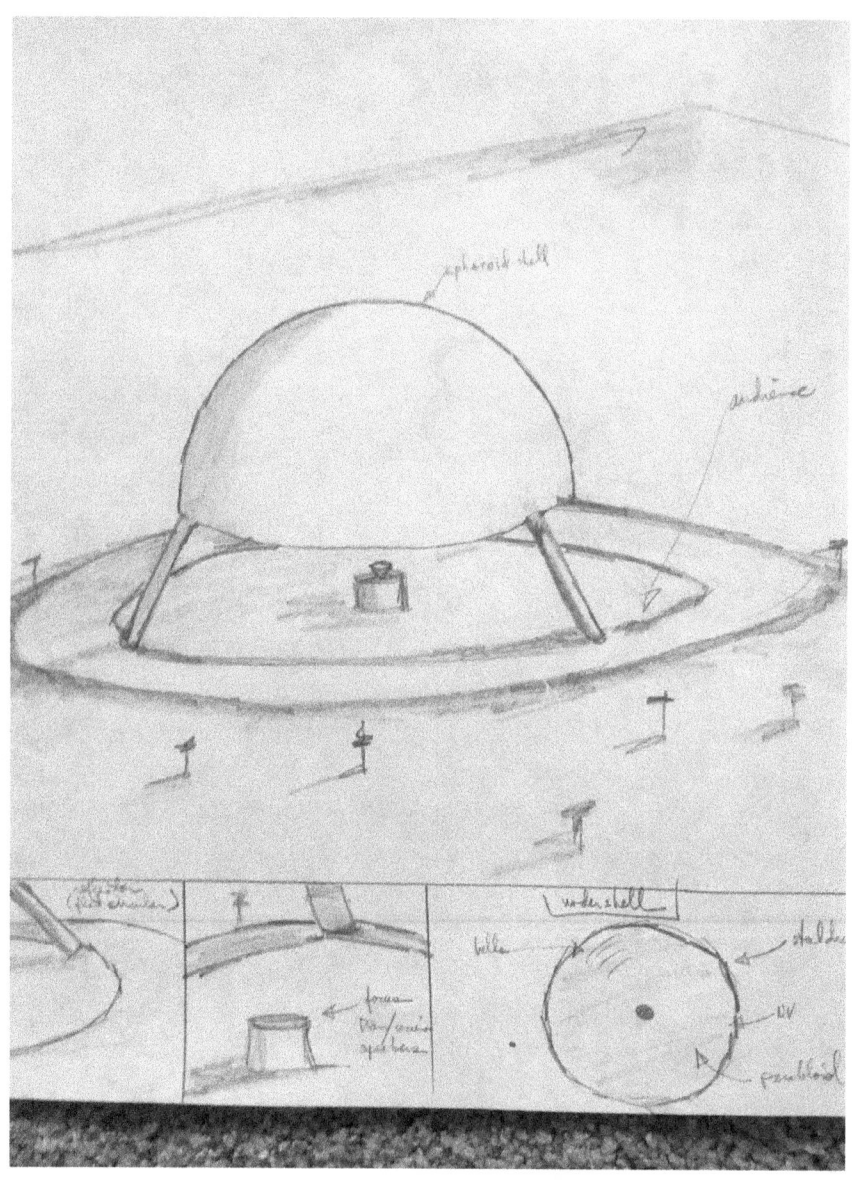

4.

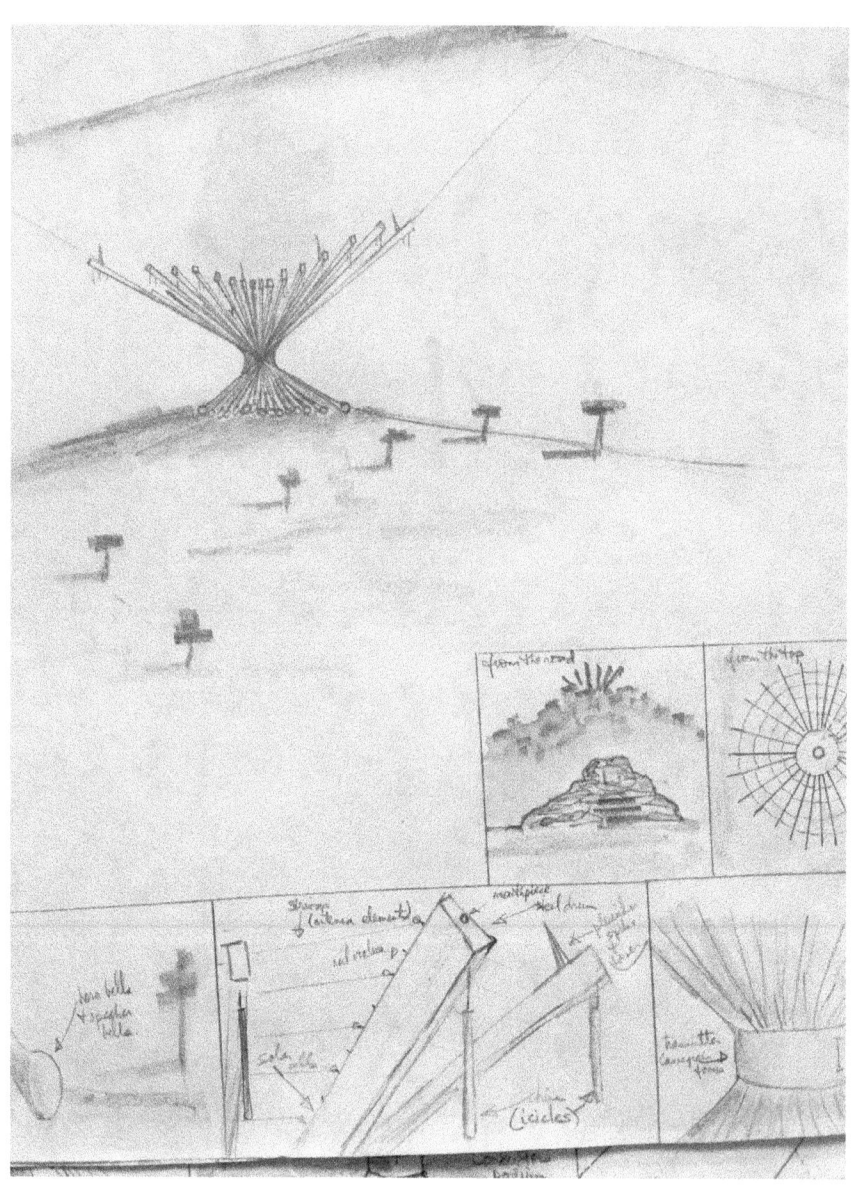

5.

6.

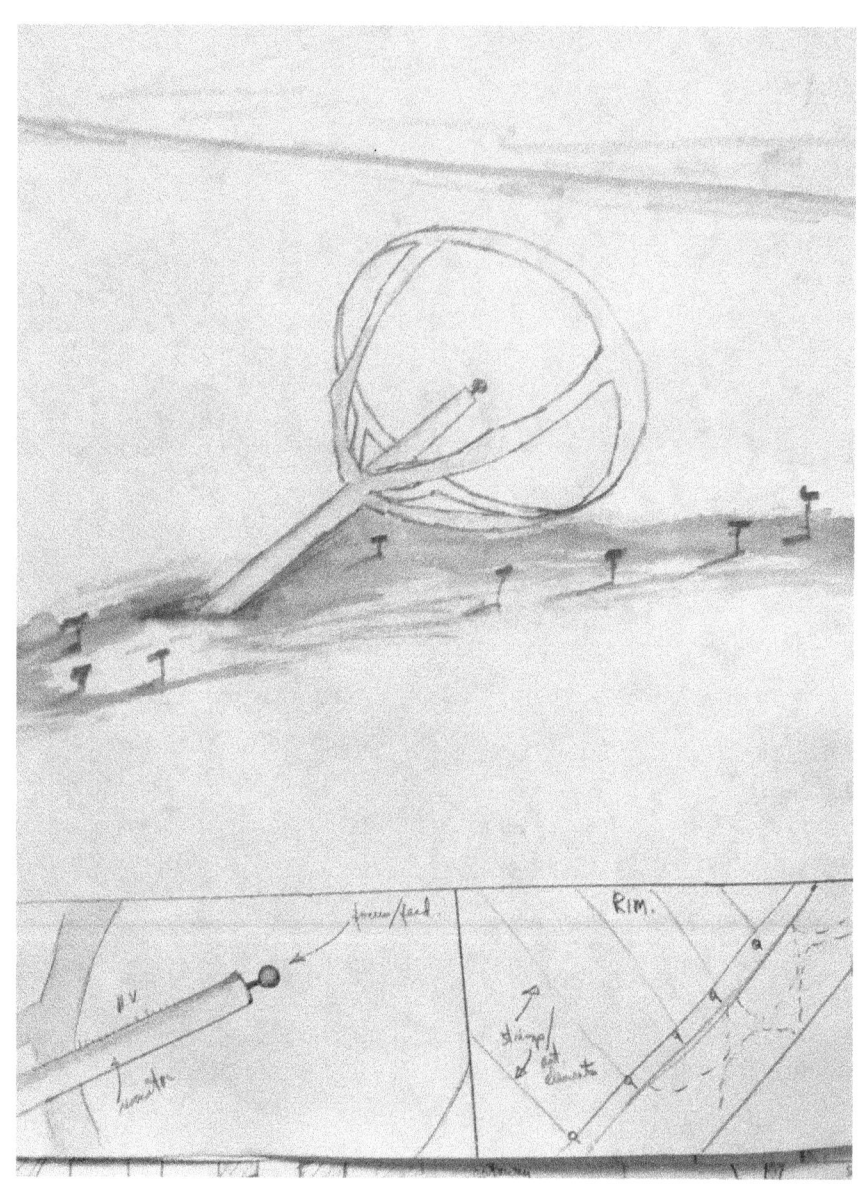

7.

8.

9.

10.

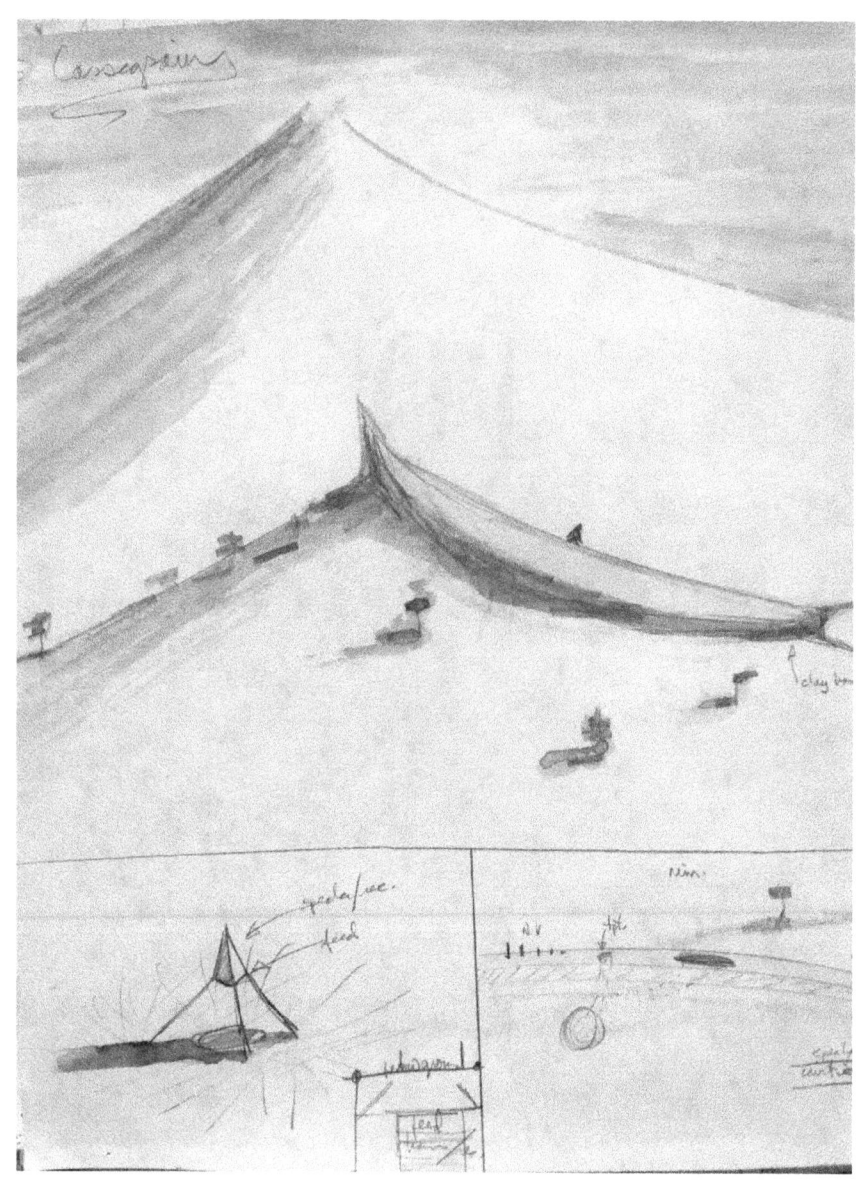

11.

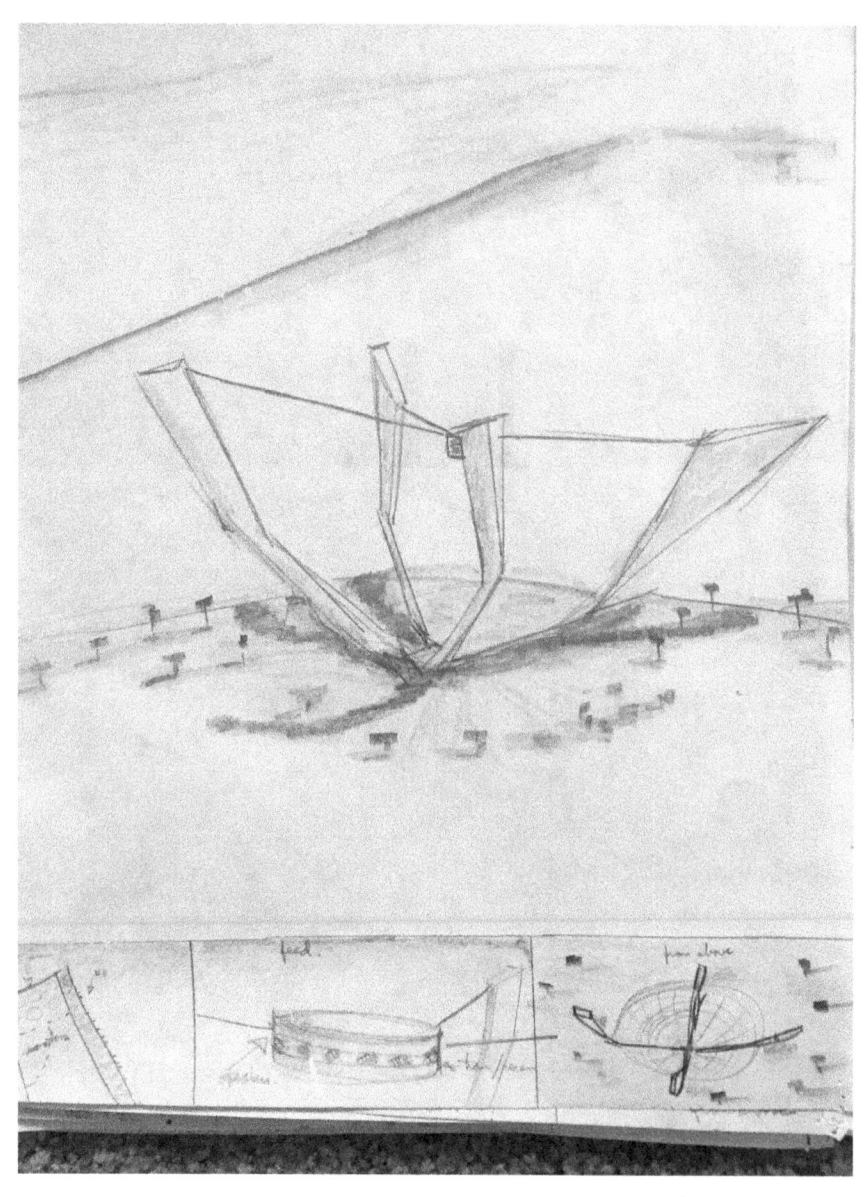

12.

13.

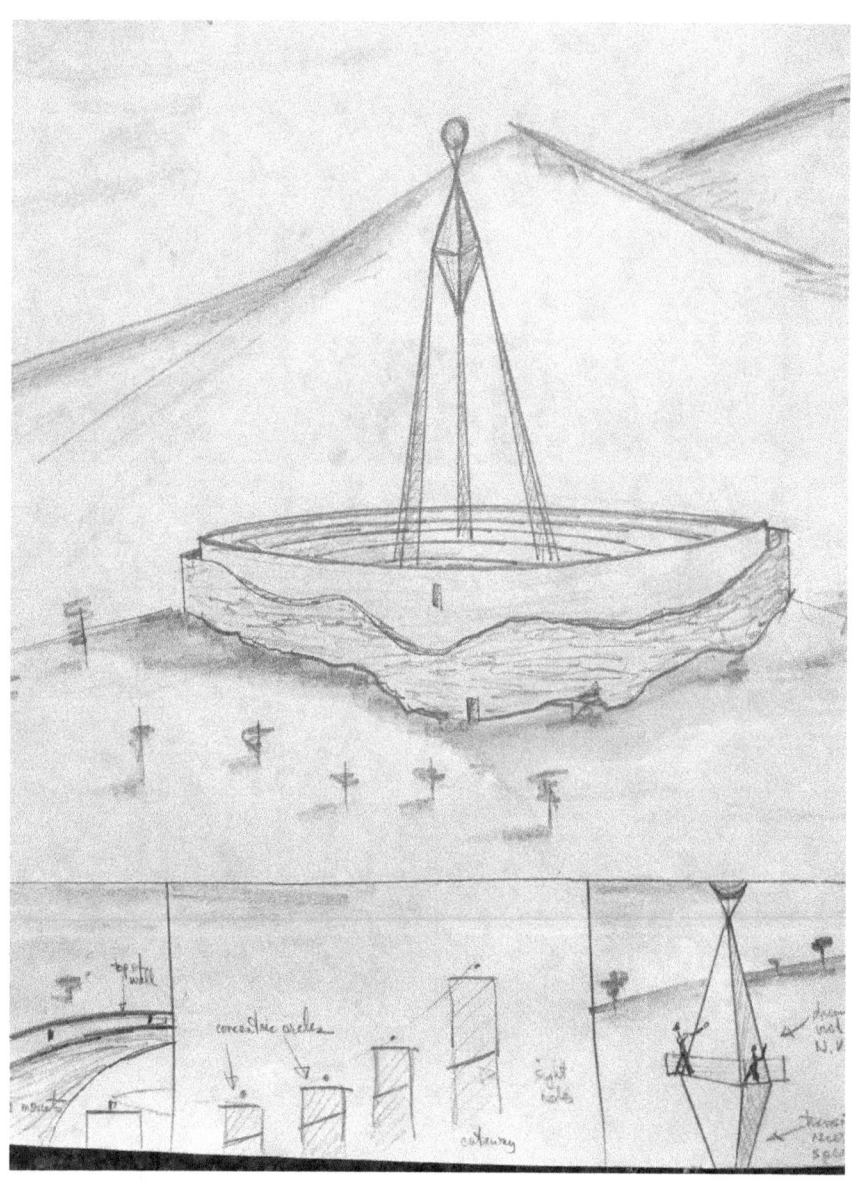

14.

15.

16.

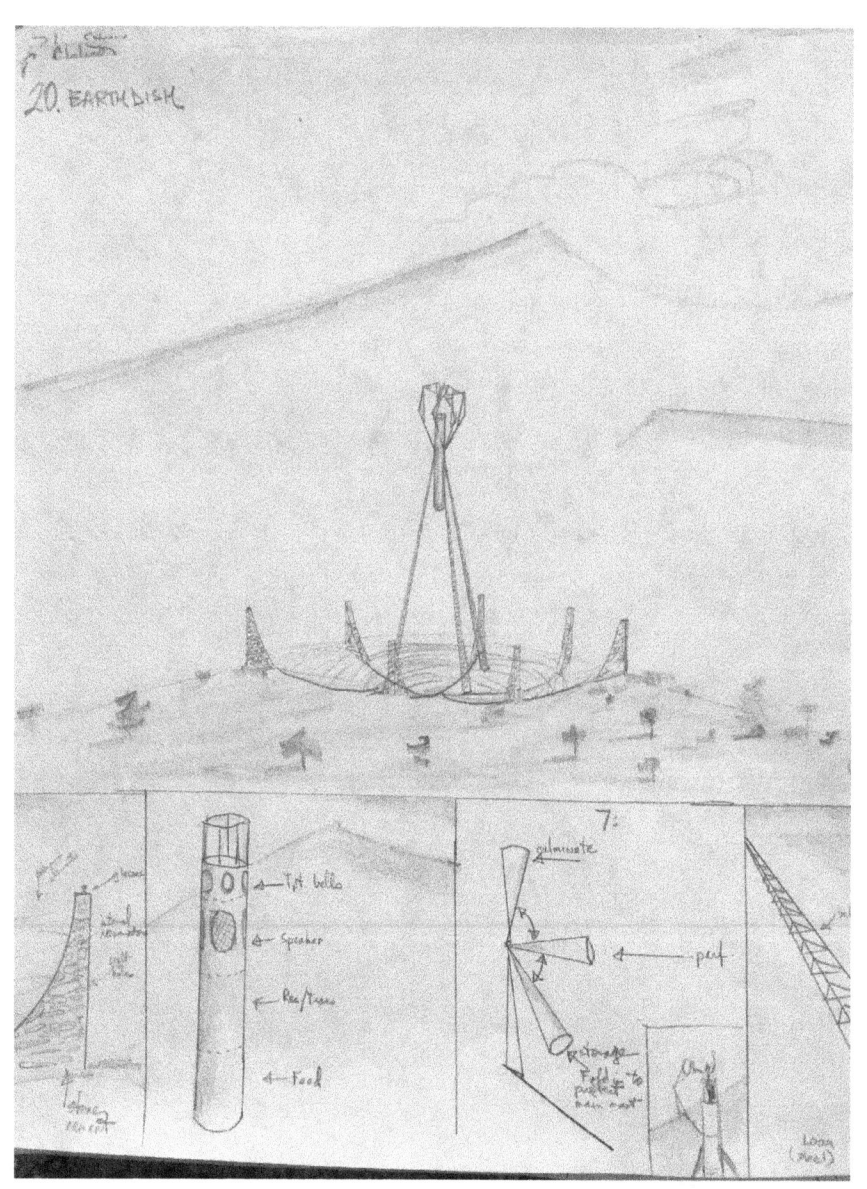

17.

18.

19.

20.

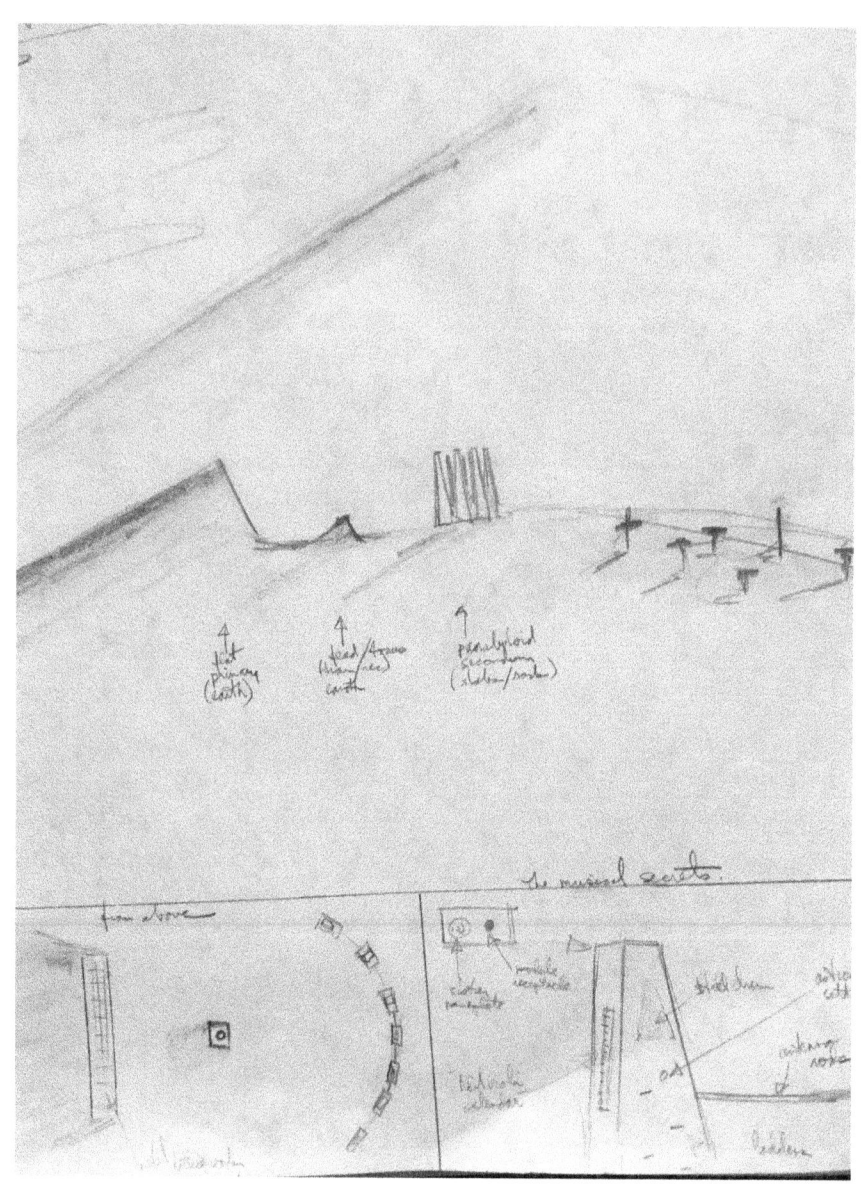

21.

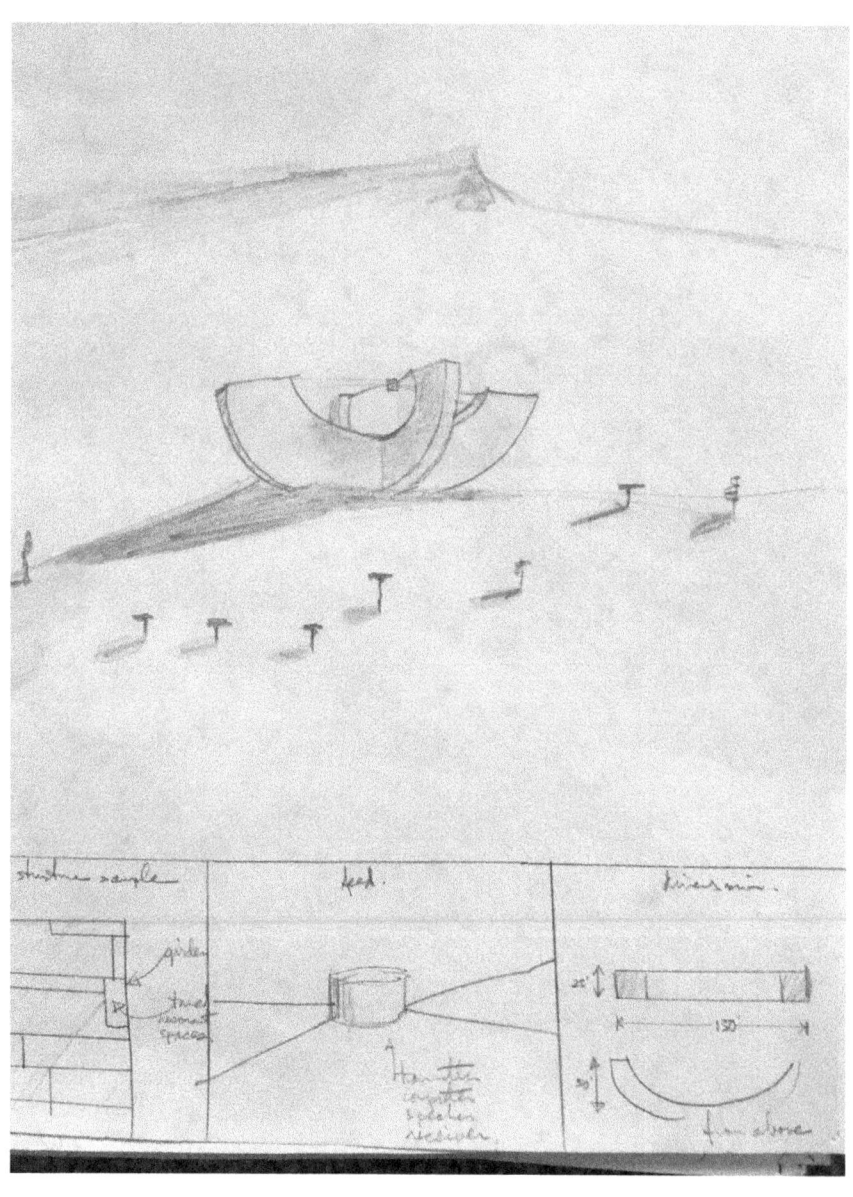

22.

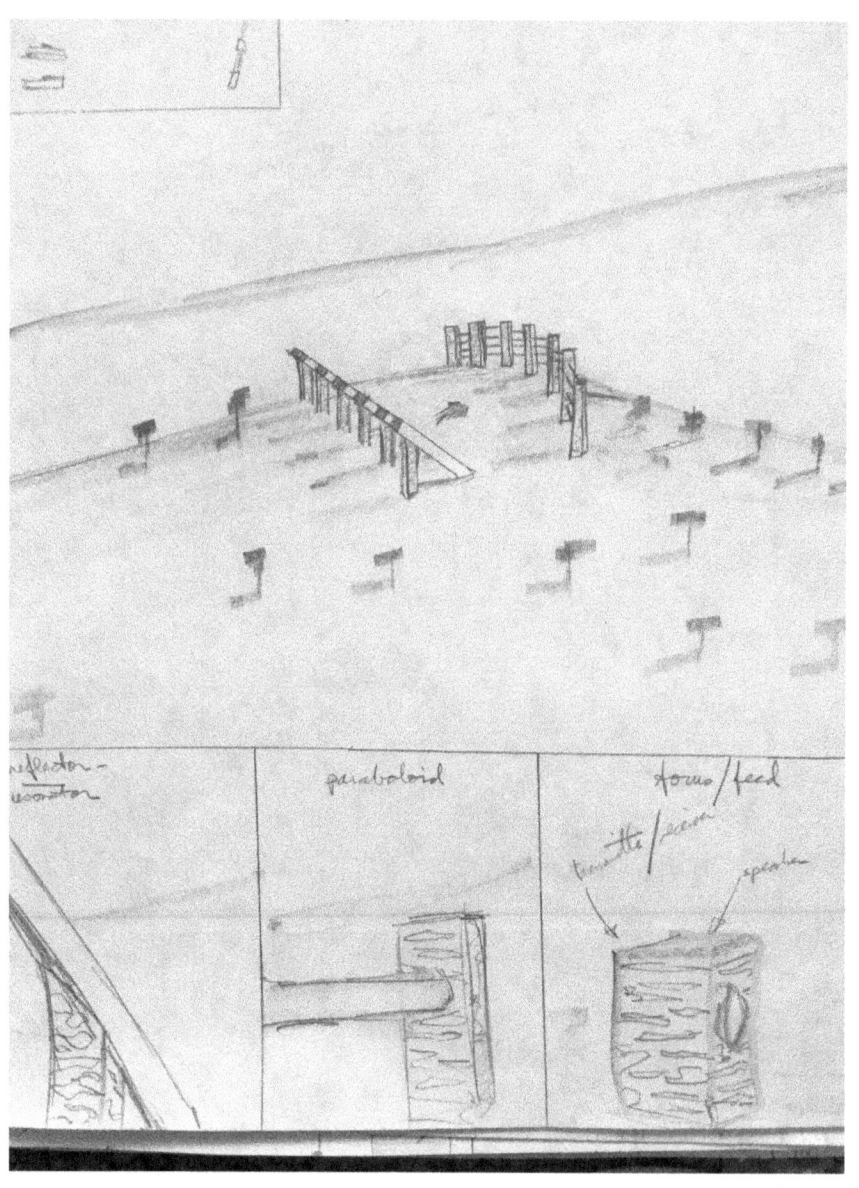

www.ingramcontent.com/pod-product-compliance
Lightning Source LLC
Chambersburg PA
CBHW081300170526
45165CB00011B/3361